Nanobiotechnology: A Complete Guide

Nanobiotechnology: A Complete Guide

Ajay Chauhan

Oxford Book Company
Jaipur ▪ New Delhi

Published By

Oxford Book Company

Regd. Off. "Bony Residency", Gate No. 2
Opp. Tilak Public School, Vishweswariya Nagar,
Gopalpura Road, Jaipur - 302018, India
Ph: 0141 - 2761280, Telefax: 2761381
email - oxfordbookjpr@gmail.com
website - oxfordbookcompany.com

Delhi Office: 102, 1st Floor, Satyam House,
4327/3, Ansari Road, Darya Ganj,
New Delhi - 110002, India
Ph: 011-45652440

Distributed By
KEYA INTERNATIONAL PRESS LLP
E-mail: keyainternationalpressllp@gmail.com

First published in India, 2026

ISBN : 978-93-5524-770-4

Typeset by Shivangi Computers, Jaipur

Printed and bound in India by Thomson Press Ltd.,
New Delhi

Contents

Preface

The phrases "bionanotechnology," "nanobiotechnology," and "nanobiology" describe the way that nanotechnology and biology interact. Since bionanotechnology and nanobiotechnology are relatively new fields, they are general words covering a variety of related technologies.

This discipline aids in highlighting the integration of several nanotechnology domains with biological research. Within the field of nanotechnology, nanobiology strengthens concepts such as nanodevices, nanoparticles, and nanoscale phenomena. By approaching biology in a technological manner, scientists are able to envision and develop systems that may be used to biological study. Nanotechnology that is biologically inspired draws inspiration from biological systems to develop new technologies. We can learn from the exquisite systems that have been produced spontaneously during ages of development. The primary goals of nanobiology are usually to apply nanotools to pertinent biological and medical issues and to improve these applications. In nanotechnology, creating new instruments for the biological and medical domains is another main goal. Frequently, the creation of new nanotools involves improving the uses of existing ones. The imaging of biological membranes, tissues, and native biomolecules is another important area of study for nanobiology researchers. The use of nanophotonics to control molecular processes in live cells and the usage of cantilever array sensors are two more areas related to nanobiology. Utilizing microbes to create functional nanoparticles has garnered a lot of attention lately. Metals may have their oxidation state altered by microorganisms. The biosynthesis of metal nanoparticles is one of the innovative applications that we may now investigate thanks to these microbial processes. The aqueous phase of microbial procedures for nanomaterial synthesis may be accomplished in a gentle and ecologically friendly manner, unlike chemical and physical approaches. In the present green bionanotechnology research towards sustainable development, this strategy has emerged as a compelling focal point. In bionanotechnology, the majority of the scientific ideas come from other disciplines. Bionanotechnology relies heavily on biochemical concepts that are used to understand the material features of biological systems since they are also utilized to develop new technologies. Mechanical characteristics, electrical and electronic, optical, thermal, biological, disease nanoscience, and computing are among the material properties and applications that are investigated in bionanoscience. The influence of bionanoscience is the conversion of biological processes into synthetic and technological applications using nanotechnology, which is accomplished through structural and mechanistic investigations of biological processes at the nanoscale. Nanotechnology provides the majority of the foundation for nanobiotechnology. Several nanotechnologies already in use serve as the direct basis for the majority of devices made for nanobiotechnological applications. When describing the overlapping interdisciplinary activities related to biosensors, especially where photonics, chemistry, biology, biophysics, nanomedicine, and engineering come together, the term "nanobiotechnology" is often used. The best way to define nanobiotechnology is as a tool that helps contemporary medicine go beyond symptom treatment to cures and biological tissue regeneration. The application of nanobiology methods by physicians has resulted in the delivery of entire cultured bladders to three American patients.

This book covers a number of novel topics, including the science of biomaterials, the creation and characterization of nanostructures, marketing strategies for nanotechnology, biological concerns, applications of nanomedicine, novel drug delivery systems, and the impact of nanotechnology on the environment and human health. This book would be helpful to researchers, instructors, and students in any scientific subject.

— ***Ajay Chauhan***

Chapter 1

Introduction

Artificial Mitochondria

In particular, the mitochondria will, at some point, fail. Increased oxygen levels in the presence of nonfunctional or partially functional mitochondria will be ineffective in restoring the tissue. However, more direct metabolic support could be provided. The direct release of ATP, coupled with selective release or absorption of critical metabolites (using the kind of selective transport system mentioned earlier), should be effective in restoring cellular function even when mitochondrial function had been compromised.

The devices restoring metabolite levels, injected into the body, should be able to operate autonomously for many hours (depending on power requirements, the storage capacity of the device and the release and uptake rates required to maintain metabolite levels). While levels of critical metabolites could be restored, other damage caused during the ischemic event would also have to be dealt with. In particular, there might have been significant free radical damage to various molecular structures within the cell, including its DNA.

If damage was significant restoring metabolite levels would be insufficient, by itself, to restore the cell to a healthy state. Various options could be pursued at this point. If the cellular condition was deteriorating (unchecked by the normal homeostatic mechanisms, which presumably would cease to function when cellular energy levels fell below a critical value), some general method of slowing further deterioration would be desirable. Cooling of the tissue, or the injection of compounds that would slow or block deteriorative reactions would be desirable. As autonomous molecular machines with externally provided power could be used to restore function, maintaining function in the tissue itself would no longer be critical. Deliberately turning off the metabolism of the cell to prevent further damage would become a feasible option.

Following some interval of reduced (or even absent) metabolic activity during which damage was repaired, tissue metabolism could be restarted again in a controlled fashion. It is clear that this approach should be able to reverse substantially greater damage than can be dealt with today.

A primary reason for this is that autonomous molecular machines using externally provided power would be able to continue operating even when the tissue itself was no longer functional. We would finally have an ability to heal injured cells, instead of simply helping injured cells to heal themselves.

Research

Advances in medical technology necessarily depend on our understanding of living systems. With the kind of devices discussed earlier, we should be able to explore and analyse living systems in greater detail than ever before considered possible.

Autonomous molecular machines, operating in the human body, could monitor levels of different compounds and store that information in internal memory. They could determine both their location and the time.

Thus, information could be gathered about changing conditions inside the body, and that information could be tied to both the location and the time of collection. Physical samples of small volumes (nano tissue samples) could likewise be taken.

These molecular machines could then be filtered out of the blood supply and the stored information (and samples) could be analysed. This would provide a picture of activities within healthy or injured tissue. This new knowledge would give us new insights and new approaches to curing the sick and healing the injured.

More dramatically, it should be feasible to take "snapshots" of tissue samples and analyse the structure down to the molecular level. First, a small tissue sample could be either fixed or frozen. Chemical fixation can be used to rapidly block most tissue changes. Ultra fast freezing of small tissue samples is an effective method of halting essentially all chemical processes and diffusion of all molecules.

Once fixed or frozen, the tissue sample could be analysed in a leisurely fashion. With nanotechnology (and indeed, to some extent with current STM and AFM technologies, though rather more expensively) it should be feasible to scan the tissue surface in molecular detail, and store that information in a computer. Once the surface had been scanned, it could be removed in a very selective and precise fashion, and scanned again.

As an example, the use of a positionally controlled carbene has been proposed for use in the synthesis of complex diamondoid structures. Such a positionally controlled carbene is highly reactive and, if positioned at an appropriate site on the surface of the tissue being analysed, would readily react with a surface molecule.

This surface molecule could then be removed. A wide variety of other "sticky" molecular tools could be brought up to the surface and allowed to react with surface molecules, which could then be removed, exposing the layers beneath. The use of a positionally controlled carbene implies that the environment in which it is used must be inert.

This requirement could be satisfied by analysing the tissue sample at very low temperature (a few Kelvins) and in a very good vacuum. Under these conditions the tissue specimen would remain stable during even a protracted analysis process.

While this process can readily be envisioned for very small structures, nanotechnology should make massive parallelism feasible. That is, a single positional device could be used at a certain speed to provide information about a certain (rather small) volume of tissue in a reasonable time.

Nanotechnology should permit the manufacture of a large number of small devices, each able to analyse a small volume. Given enough such devices operating in parallel, larger volumes could be analysed and the information from many individual devices integrated to provide a coherent picture of the larger whole. Effective use of this option will require massive computational power—which will also be made feasible with nanotechnology.

Estimates of the computational power that should be provided by nanotechnology exceed 10^{24} logic operations per second for a single desktop computer. This amount of raw computational power should make control of a large number of parallel devices feasible, and should permit integration and analysis of the information so obtained. In short, tissue samples could be "frozen" (either literally by ultrafast cooling or figuratively by chemical fixation) and the entire resulting tissue sample could be analysed down to the level of individual molecules. The information so obtained could be processed by computers able to handle the flood of data produced.

The resulting "snapshots" will provide us with an instantaneous look at metabolic and cellular activities across even relatively large volumes of tissue. Such an ability should revolutionize our understanding of the complex processes that take place in living systems. The possibility of truly revolutionary advances in our medical abilities has also created renewed interest in cryonics. The abilities discussed here might well take years or decades to develop.

It is quite natural to ask: "When might we see these systems actually used?" The scientifically correct answer is, of course, "We don't know." That said, it is worth noting that if progress in computer hardware continues as the trend lines of the last 50 years suggest, we should have some form of molecular manufacturing in the 2010 to 2020 time frame.

After this, the medical applications will require some additional time to develop. The remarkably steady trend lines in computer hardware, however, give a false sense that there is a "schedule" and that developments will spontaneously happen at their appointed time.

Control to Chemical Synthesis

Manufactured products are made from atoms. The properties of those products depend on how those atoms are arranged. Viewed from the molecular level today's macroscopic manufacturing methods are crude and imprecise. Casting, milling, welding and all the other traditional manufacturing methods spray atoms about in great statistical herds. Even lithography (which already lets us put millions of transistors on a chip no bigger than your fingernail) is fundamentally statistical and random.

Exactly how many dopant atoms are in a single transistor and exactly where each individual dopant atom is located is neither specified nor known: if we have roughly the right number in roughly the right place, we can make a working transistor. For today, that is good enough.

The exception is chemistry. Large high purity crystals can have almost every atom in the right place. So, too, can many long polymers. The structures of proteins with hundreds and even thousands of amino acids can be specified down to the last atom. Most dramatically (and fortunately for us!) DNA strands with many tens of millions of bases can be copied with almost perfect accuracy.

And it seems that almost any small molecule (with perhaps several dozens of atoms) can be synthesized, if only we have the skill and patience. Yet the laws of physics and chemistry in principle permit arranging and rearranging the elements in so many combinations and permutations that all of our manufacturing skills and all of our chemical skills barely suffice to scratch the surface of what is possible.

The Utility of Diamond

Almost any manufactured product could be improved, often by several orders of magnitude, if we could precisely control its structure at the molecular level. We often want our products to be light and strong. Diamond is light and strong: the strength-to-weight ratio of diamond is over 50 times that of steel.

Yet we do not today have diamond spars in airplanes nor diamond hulls for rockets. Today we can't economically make diamond. Even if we could, simple diamond crystals can shatter. We'd have to modify the structure to make it tough and shatter proof: perhaps diamond fibres. While easily done in principle, we can't do this in practice today. Great strength and light weight are not the exclusive province of diamond: graphite can be stronger. And if we consider the many ways in which carbon atoms can be arranged and rearranged, then it's obvious that there are a host of other possibilities.

Yet all share a common problem: we can't yet economically make them in the exact shapes that we want. Great strength is only one property that we prize highly: when we make computers we are more concerned by electrical properties. Here, too, diamond excels.

Today's computers are made of semiconductors, and the semiconductor of choice is silicon. This is not because silicon is the ideal semiconductor from which to make computers, but because we know how to make devices from it.

The computer industry has strong opinions about what makes a good logic device and what makes a good computer and diamond will let us make better computers than silicon. Diamond has a wider bandgap, hencc electrical devices will work at higher temperatures.

It has greater thermal conductivity, so devices can be more easily cooled. It has a greater breakdown field, hence devices can be smaller. It has higher electron and hole mobility which, when combined with higher electric fields, will result in higher speed.

But again, we see no diamond computers, just as we see no diamond airplanes: we can't economically manufacture them yet. Large pure crystals of silicon can be made relatively easily, but large pure crystals of diamond are scarce.

We can etch the silicon surface and add dopants with a precision measured in tenths of microns, while the corresponding steps for diamond are more difficult. Not more difficult in principle: just more difficult today.

Making computers highlights another problem. It's not enough to make a pure crystal, it must also have an extremely precise and complex pattern of impurities. The exact location of the dopant atoms in the semiconductor lattice controls how devices function and where signals can propagate.

Local order is crucial to make each device work, but long range complex order is crucial to make the computer as a whole work. While we can make some things today that are highly precise and have simple long range order (*e.g.*, crystals), it is the requirement for complex long range order that prevents us from making computers of the kind we'd like to make.

While it's plausible we could make high density memory from crystals and perhaps some types of cellular automatons, we couldn't make anything that resembled the computers on the market today.

Today's high speed semiconductor-based digital computers (like the 80486 or the Pentium) have millions of logic elements wired together in complex and highly idiosyncratic patterns.

This is well beyond the capabilities of crystal growth or bio-polymer synthesis. It will require a fundamentally new manufacturing technology: molecular manufacturing.

Today, there is a gap in our synthetic abilities: we can make complex mechanical machinery and electronic devices (including computers, which have millions of transistors), but we can't make such devices with the precision with which the chemist can synthesize a crystal, a bio-polymer, or a relatively small molecule. With chemistry we can make precise molecular structures and compounds, but we haven't been able to scale up that success to molecular computers (and other macroscopic products as precise as molecules).

Molecular manufacturing will, by definition, let us economically manufacture almost any specified structure that is consistent with the laws of chemistry and physics. To simplify the problem somewhat we can narrow our focus to structures that resemble diamond in a broad sense: the diamondoid structures as defined by Drexler.

This class includes (among other things) diamond crystals of arbitrary shape but with stably terminated surfaces (*e.g.*, hydrogenated (111) or the like) and with impurities at precise locations in the diamond lattice (*e.g.*, substitutional boron).

Our objective is to manufacture particular diamondoid structures once the location and type of every atom has been specified by design.

Computer Industry

The attraction of molecular manufacturing for the computer industry should be clear. It should let us make computers at a manufacturing cost of less than a dollar per pound, operating at frequencies of tens of gigahertz or more, with linear dimensions for a single device of roughly 10 nanometers, high reliability, and energy dissipation (using conventional methods) of roughly 10^–18 joules per logic operation.

If we make thermodynamically reversible computers (which the author and others have recently shown can be made from conventional electronic devices, *e.g.*, CMOS) then the energy dissipation per logic operation can be reduced to well below kT at T = 300 Kelvins (well below 10^–21 joules).

The computer industry is spending billions of dollars to make better computers. It is widely acknowledged within the industry that lithography is approaching its limits. Articles like *The Future of the Transistor*, *Miniaturization of Electronics and its Limits* and *Outlook for VLSI: Will the Balloon Burst?* Quite clearly show that conventional lithography will run out of steam (in perhaps a decade, though there is less agreement about the exact time frame).

There is already interest in molecular logic devices and that interest will increase sharply as improvements in conventional manufacturing methods become increasingly difficult. However, any new proposal for manufacturing molecular computers will be weighed against (at least) the criteria mentioned above.

If it cannot easily beat conventional methods after they have been pushed to their uttermost limits, then it will be rejected.

The computer industry will soon be pouring vast sums into research aimed at molecular computing, but the great bulk of funding will go towards well thought out proposals that offer a realistic possibility of substantially exceeding the performance of the ultimately evolved silicon VLSI technology that we expect to develop over the next decade.

If you can't beat tomorrow's mainstream computers, you might as well not try. For this and many other reasons the class of diamondoid structures is a reasonable one to consider. The problem of building a diamondoid electronic computer captures many of the fundamental issues in molecular manufacturing, and poses clearly the issue of building large structures that cannot be made by regular repetition of some substructure (*e.g.*, the unit cell of a crystal or the monomeric unit in a bio-polymer).

This brings us to a core issue in molecular manufacturing: how do we synthesize such things? Today, we can synthesize diamond at low pressure and low temperature by using CVD (Chemical Vapour Deposition) methods.

Diamond CVD growth involves highly reactive species (radicals, carbenes, etc.) in a gas over the growing diamond surface that bombard and react with that surface at random. Because reaction sites are random, growth of many defect structures occurs (dislocations, etc.) as well as the desired perfect diamond structure.

Two fundamental mechanisms in the growth process include:

- Abstraction of hydrogens from the diamond surface leaving behind reactive sites (dangling bonds, radicals) and
- Interaction of carbon species (both reactive (CH2, CH3, etc.) as well as relatively unreactive species (C2H2)) with the surface, thus depositing carbon.

If we are to synthesize diamondoid structures it is plausible that we begin our search for the basic reaction steps involved in this synthesis by looking at existing reactions that occur in the CVD growth of diamond. The use of a reactive gas in the synthesis process, however, would seem to defeat any hope of making precisely patterned diamondoid structures, for the gas will interact with the growing surface at random.

Positional Control is Fundamental

Here, we introduce the fundamental concept of molecular manufacturing: positional control over the site of reactions. To take a specific example we consider site specific hydrogen abstraction from the diamond (111) surface.

The ability to remove specific hydrogen atoms from the surface of the diamondoid work piece under construction is likely to be a fundamental unit operation in any attempt to make atomically precise diamondoid structures. Hydrogen abstraction during CVD diamond growth typically involves a radical reaction between atomic H from the gas with H bonded to carbon on the surface producing H2.

It is unclear how to make this process site specific. However, there are other structures with a high affinity for hydrogen which offer greater possibilities for positional control. In particular, the propynyl radical C3H3 has a great affinity for hydrogen. Further, this radical has the very useful property that it has two ends: one end is a highly reactive radical while the other end is a stable sp3 carbon. Thus, we could synthesize a larger molecule with the propynyl radical at its end. The larger molecule would be held at the tip of a positional device.

The positional device would provide control over the orientation and position of this hydrogen abstraction tool (*e.g.*, a six degrees of freedom manipulator) and thus control the site of abstraction by controlling the position of the tool.

Ab initio quantum chemical analysis of the abstraction of hydrogen from isobutane using an ethynyl abstraction tool supports the idea that the barrier to this reaction is zero. The reaction will proceed rapidly and, because of the large exothermicity, irreversibly.

Calculated barriers for abstraction from several other molecules were also small, suggesting that this hydrogen abstraction tool could be used to abstract hydrogen from a wide range of different molecules.

The site specific abstraction of hydrogen illustrates the core concept in molecular manufacturing: selecting the reaction site by controlling the position and orientation of the reactants. The (relatively stiff) diamondoid workpiece is held in place, while a tool (in our example, a hydrogen abstraction tool) is positioned using a rather conventional (if also rather small) robotic arm.

The ideas of using tools, controlling the position of those tools with a general purpose manipulator, and building complex structures by putting together components using those positionally controlled tools are rather common and even mundane at the macroscopic level.

At the molecular level, they are new and almost shocking: yet it is simply mapping onto the molecular world the concepts and ideas that have proven so useful and powerful in macroscopic manufacturing. By adding positional control we should be able to develop a method of molecular manufacturing which combines the best features of both conventional macroscopic manufacturing and chemical synthesis.

Molecular Tools

If we are to grow diamond, we must also have carbon deposition tools. Drexler has suggested the use of positionally controlled carbenes and alkynes and proposed reaction pathways and surface structures where these tools would apply. In both cases, the tools are positioned at a precise point on the growing diamondoid structure and are used to deposit one or more carbon atoms at a desired location.

These deposition reactions parallel proposals in the CVD literature except for the addition of positional control (*e.g.*, at least one portion of the moiety must be part of an extended "handle" which can be held by a positional device). The broad range of possible tools coupled with the great power of ab initio computational chemistry should let us define and verify a complete set of molecular tools capable of synthesizing essentially any diamondoid structure.

Tool Use

For such tools to be usable in a system context we must satisfy certain constraints. First and foremost, we must have a device capable of positioning the tool to within something like an atomic diameter. On the diamond (111) surface, the distance between adjacent hydrogens is about 2.5 Angstroms.

Thus, positional accuracy of 1 to 2 angstroms for the hydrogen abstraction tool is required to prevent abstraction of the wrong hydrogen. Second, because the tools can be highly reactive, we require an inert environment. A simple inert environment is vacuum.

Compressed helium or some other inert gas would also work. Third, because it is the relative tool-workpiece position that must be controlled, the workpiece under construction must be relatively rigid (*e.g.*, not subject to vibrational motions that would exceed about an angstrom). Fourth and last, we must have some way of generating the sometimes highly reactive tools (*e.g.*, we need to define a precursor to the hydrogen abstraction tool, as well as precursors to the other tools). In some sense, the analysis that we will now pursue is similar in type to retrosynthetic analysis. We start with the final product that we wish to build (a macroscopic diamondoid computer, for example), and then consider the possible predecessor structures which would yield the final product in one step.

Then we consider the predecessor structures to those structures, and so on. We extend retrosynthetic analysis beyond its traditional bounds, but the general concept remains the same: given the finished product we deduce the possible ways in which it could have been constructed. This approach has been called "backwards chaining" by Drexler.

Work with SPMs (Scanning Probe Microscopes) clearly show that it is possible to achieve positional accuracies of a small fraction of an angstrom. Small (~0.1 microns) diamondoid "arms" or positional devices with similar positional accuracy are in principle quite feasible.

The field of robotics provides a broad range of designs for positional devices which are largely scale independent. Shrinking these designs to submicron size is conceptually straightforwards. A factor of crucial importance in the design of molecular-scale positional devices is the accuracy with which the tip can be positioned, particularly in the face of thermal noise.

While atomically precise bearings and joints will not suffer from chatter, backlash, wear, tooth-to-tooth errors and other sources of inaccuracy caused by imprecise manufacturing, they will still suffer from positional errors caused by thermal noise. To control this source of error, it is essential that the robotic arm be very stiff, and so the use of stiff materials is desirable.

The Young's modulus for diamond is about 10^12 Pascals (very stiff), and back-of-the-envelope calculations show that a hollow cylinder of such material that is perhaps 100 nanometers long and 30 nanometers in diameter should have a positional accuracy at the tip, in the face of thermal noise at room temperature, of a small fraction of an atomic diameter.

A more detailed design and analysis of a jointed tubular robotic arm taking into account the bending and rocking motions of joints in the arm further supports this conclusion. Alternatives to the simple robotic arm are available which might be more attractive.

Other Requirements for Tool Use

Creating an inert environment also presents no fundamental problems: high quality vacuums are common in laboratories today.

If our objective is to have a very small very high quality vacuum, then a relatively thin wall of diamondoid material could be used as a barrier to keep a volume which was a modest fraction of a cubic micron free of any contaminants. If the volume were initially constructed free of contaminants then such a barrier would keep the inside free of any contaminants with high probability.

Because we are building diamondoid structures, they will be very stiff. As a consequence, it is relatively easy to meet the requirement that the objects that are being manufactured must themselves be stiff. Finally, generation of "activated" tools from relatively stable precursors can be done by a variety of methods. Because we are assuming an environment in which we have positional control we can use particularly simple precursors.

Transport Across a Barrier

Having introduced a diamondoid barrier to keep unwanted contaminants out (much as the bacterial wall allows bacteria to maintain an appropriate internal environment in the face of a fluctuating external environment) we must now solve the problem of getting desired raw materials through the barrier. We might, for example, wish to transport the hydrogen abstraction tool precursor across the diamondoid barrier.

After use we will also need to eject the spent tool. Several ways to solve this problem are feasible. A proposal by Drexler is to use a rotor embedded in the diamondoid wall which moves binding sites from the outside of the wall to the inside of the wall.

By modulating the affinity of the binding site so that it will have high affinity for the desired molecule outside the barrier and low affinity inside the barrier, efficient transport across the barrier can be achieved.

The desired molecule will bind to the binding site when it is outside the barrier, the binding site will be rotated to the inside of the barrier and the binding affinity reduced (in the illustrated proposal by mechanically pushing a rod into the binding site, thus physically precluding occupancy), and the molecule will be released on the inside of the barrier.

The result is to increase the concentration of the desired molecule. A few stages of such a filtration system can achieve extremely high purities. The final stage, rather than ejecting the molecule into a liquid, would deliver the molecule into the inert internal environment in a well defined orientation where it could be further processed.

Control Signals

Finally, we will need a source of control signals for our molecular arm. One general approach would be to use a molecular computer.

We will not consider a particular design for a molecular computer here, it is sufficient to note that many proposals for molecular computation have been considered in the literature and it is generally expected that some type of very small computational device will be feasible in a few decades.

This completes our (all too brief) outline of a small device able to manufacture a broad class of diamondoid materials. Basically, the design is driven by the desire to provide the environment needed to synthesize diamond and diamondoid materials using the kinds of reactions that occur naturally during CVD growth of diamond.

The device is itself made from diamondoid materials, which means that one such device can manufacture a second such device.

This ability to self replicate is crucial in achieving low manufacturing costs. As the reader might appreciate, the design and construction of one such general purpose device might well prove to be a time consuming and expensive undertaking. This cost cannot be justified unless the resulting device has great value.

If the device can self replicate then the successful design and development of one such device can be used to build an entirely new manufacturing technology. The manufacturing costs for the second, third, fourth.... 10^10.... etc. devices will consist largely of the raw materials and energy costs. Thus, a very large R&D cost can (if necessary) be justified.

The major driving force in the previous design was the desire to approach fundamental limits (strength, stiffness, thermal conductivity, electrical characteristics, etc.) in the manufactured products. This in turn implied the use of diamondoid materials, which, when coupled with the known chemistry of CVD diamond synthesis resulted in a high vacuum system with highly reactive intermediates.

Relaxing the materials requirements gives us a *much* wider range of possible structures. In particular, we can consider what is sometimes called "brick" or "building block" based nanotechnology. In this approach we first design a set of molecular building blocks, and then assemble the building blocks by the use of positional control.

The ribosome can be viewed as the prototype for this approach. The building blocks are amino acids, and they are linked together by the ribosome to form proteins.

Our approach differs in two principle respects: first, we add positional and orientational control over the building blocks in three dimensions, while the ribosome can only build structures that are fundamentally one dimensional (relying on linear structures that spontaneously fold into a particular shape to achieve a degree of control in three dimensions).

Second, rather than using relatively floppy polymers, we prefer relatively rigid bricks that can be bonded to each other in a stiff three-dimensional framework. This greatly simplifies the system. Indeed, with brick-based nanotechnology one can relatively easily envision the synthesis of a set of bricks that can, with the addition of positional control, be assembled into a wide range of structures with the stiffness of (say) wood.

A conceptually simple and relatively near-term way to achieve some degree of positional control would be to use conventional SPMs. The tips used in current SPM's are usually quite crude, and even when it is possible to make a very fine tip the range of possible structures is very limited (tungsten or some other simple material is generally all that's available).

The design of tips for the SPM that incorporate individual molecules specifically synthesized for the purpose is a likely next step, and one that seems essential if we are to make progress in using SPM's to guide chemical reactions in a selective way. Such a "molecular manipulator" should be within reach of today's experimental technology.

While the Young's modulus of the things it could make would be substantially inferior to that of diamond, this can be compensated by making them bigger. Scaling laws are such that increasing the size of an object by a factor of 10 also increases its stiffness by a factor of 10, and so reduces the positional inaccuracy from thermal vibration by a factor of 10.

If we wish to build a molecular "arm" out of bricks, then to achieve the same positional accuracy as with a diamondoid arm we will have to make the arm bigger—perhaps several tenths of a micron or more—but the basic design concepts that were discussed previously for use with a diamondoid arm still hold.

We can still position the individual bricks accurately both in position and orientation, we can still build larger structures by putting together many bricks, we can still control the synthesis process by positional means, we can still make a positional device from bricks, etc.

Why pursue such an approach when it can only make relatively inferior materials?

There are three primary reasons:

1. It's easier to do
2. It could still make many things that are very valuable by today's standards and
3. Such systems could be used to make better systems (*e.g.*, diamondoid systems).

As the reader will appreciate, there are many possible candidates for bricks. Many researchers are already considering the design of molecular building blocks, although in most cases they do not consider positional control. Krummenacker discusses several of the issues surrounding the design of molecular bricks intended for assembly via positional control.

Adding positional control makes the synthesis of a broad range of molecular structures feasible, but at the same time requires the design and synthesis of a mechanism able to provide such control.

The obvious first step is to provide positional control using relatively modest extensions to today's SPMs. The most significant addition is the inclusion of a molecular tip, for today's SPMs typically have a limited range of possible tip structures and are often imprecise at the molecular scale.

Current Findings

Manufacturing

Molecular manufacturing is the use of programmable chemistry to build exponential manufacturing systems and high-performance products. There are several ways this can be achieved, each with its own benefits and drawbacks.

This technology is coming soon—almost certainly within 20 years, and perhaps in less than a decade. When it arrives, it will come quickly.

Molecular manufacturing can be built into a self-contained, tabletop factory that makes cheap products efficiently at molecular scale. The time from the first assembler to a flood of powerful and complex products may be less than a year. The potential benefits of such a technology are immense. Unfortunately, the risks are also immense.

Products Value

Even a primitive diamond-building nanofactory can create products vastly more powerful than today's versions. Electrical power can be converted to motion, and *vice-versa*, with one-tenth the power loss and about 108(100,000,000) times more compactly. Computers can be a billion times smaller and use a million times less power.

Materials can be about 100 times stronger than steel. This means that most human-scale products would consist almost entirely of empty space, reducing material requirements and cost. Most of the rest of the product would be structural, easy to design. Even the simplest products could be software-controlled at no extra hardware cost.

Manufacturing of prototypes would be quite rapid-a few minutes to a few hours. Because manufacturing and prototyping are the same process, a successful prototype design could immediately be distributed for widespread use. A designer working with a few basic predesigned blocks could design, build, and test a simple product in less than a day.

Products with complex interfaces to humans or to their surroundings-information appliances, automobiles, aerospace hardware, medical devices-would be limited by the time required to develop their software and test their functionality.

However, in some fields the high time and money cost of manufacture slows other parts of the development cycle; this effect would disappear. An explosion of new, useful products could rapidly follow the widespread availability of a nanofactory.

Advantage of Ced Nanotech

Manufacturing with nanotechnology can solve many of the world's current problems. Water shortage is a serious and growing problem. Most water is used for industry and agriculture; both of these requirements would be greatly reduced by products made by molecular manufacturing.

Infectious disease is a continuing scourge in many parts of the world. Simple products like pipes, filters, and mosquito nets can greatly reduce this problem. Information and communication are valuable, but lacking in many places. Computers and display devices would become stunningly cheap. Electrical power is still not available in many areas.

The efficient, cheap building of light, strong structures, electrical equipment, and power storage devices would allow the use of solar thermal power as a primary and abundant energy source. Environmental degradation is a serious problem worldwide. High-tech products can allow people to live with much less environmental impact. Many areas of the world cannot rapidly bootstrap a 20th century manufacturing infrastructure.

Molecular manufacturing can be self-contained and clean; a single packing crate or suitcase could contain all equipment required for a village-scale industrial revolution. Finally, MNT will provide cheap and advanced equipment for medical research and health care, making improved medicine widely available. Much social unrest can be traced directly to material poverty, ill health, and ignorance. MNT can contribute to great reductions in all of these problems, and in the associated human suffering.

Advanced Nanotech

Molecular nanotechnology will be a significant breakthrough, comparable perhaps to the Industrial Revolution-but compressed into a few years. This has the potential to disrupt many aspects of society and politics.

The power of the technology may cause two competing nations to enter a disruptive and unstable arms race. Weapons and surveillance devices could be made small, cheap, powerful, and very numerous. Cheap manufacturing and duplication of designs could lead to economic upheaval.

Overuse of inexpensive products could cause widespread environmental damage. Attempts to control these and other risks may lead to abusive restrictions, or create demand for a black market that would be very risky and almost impossible to stop; small nanofactories will be very easy to smuggle, and fully dangerous. There are numerous severe risks-including several different kinds of risk-that cannot all be prevented with the same approach. Simple, one-track solutions cannot work.

Simplistic Regulation

Molecular nanotechnology manufacturing creates several severe risks, and each risk tempts a simple and extreme solution. However, a patchwork of extreme solutions will be both destructive and ineffective. For example, Bill Joyand others have proposed halting nanotechnology research entirely.

This would not actually work; instead, it would relocate the research to less responsible venues. The risks might be delayed by a few years, but would be far worse when they appeared because the technology would be even less controllable.

To take another example, economic upheaval might be prevented by strict commercial licensing of all uses of the technology. This has two problems. First, digital protection schemes for commercial products have often proved quite easy to crack.

Second, if the technology is so restricted that it cannot disrupt existing economic systems, continuing poverty will kill millions of people each year, fueling backlash, social unrest, espionage, and independent development. Each risk must be reduced by some means that does not exacerbate others.

Approaches

Once molecular manufacturing is developed, it will have to be administered. There are several approaches that might help. CRN is not advocating any of these approaches at this point; we don't know enough about how the technology will be developed or in what context.

We also have to point out that we don't think any one approach will be enough. Any effective program will require a balance of several different kinds of administration. Some possibilities include built-in technical restrictions in personal nanofactories; intellectual property reform; and international cooperation or monitoring of various kinds. Despite the difficulties and complexities, we believe that a solution can be found to preserve most of the potential benefits while avoiding the most severe risks.

Chapter 2

Nanotechnology of Cloning

Organism Cloning

Organism cloning (also called reproductive cloning) refers to the procedure of creating a new multicellular organism, genetically identical to another. In essence this form of cloning is an asexual method of reproduction, where fertilization or inter-gamete contact does not take place. Asexual reproduction is a naturally occurring phenomenon in many species, including most plants and some insects. Scientists have made some major achievements with cloning, including the asexual reproduction of sheep and cows. There is a lot of ethical debate over whether or not cloning should be used. However, cloning, or asexual propagation, has been common practice in the horticultural world for hundreds of years.

Horticultural

The term *clone* is used in horticulture to refer to descendants of a single plant which were produced by vegetative reproduction or apomixis. Many horticultural plant cultivars are clones, having been derived from a single individual, multiplied by some process other than sexual reproduction.

As an example, some European cultivars of grapes represent clones that have been propagated for over two millennia. Other examples are potato and banana. Grafting can be regarded as cloning, since all the shoots and branches coming from the graft are genetically a clone of a single individual, but this particular kind of cloning has not come under ethical scrutiny and is generally treated as an entirely different kind of operation. Many trees, shrubs, vines, ferns and other herbaceous perennials form clonal colonies. Parts of a large clonal colony often become detached from the parent, termed fragmentation, to form separate individuals. Some plants e.g. dandelion also form seeds asexually, termed apomixis, resulting in clonal populations of genetically identical individuals.

Parthenogenesis

Clonal derivation exists in nature in some animal species and is referred to as parthenogenesis (reproduction of an organism by itself without a mate). This is an asexual form of reproduction that is only found in females of some insects, crustaceans and lizards. The growth and development occurs without fertilization by a male. In plants, parthenogenesis means the development of an embryo from an unfertilized egg cell, and is a component process of apomixis. In species that use the XY sex-determination system, the offspring will always be female. An example is the "Little Fire Ant" (*Wasmannia auropunctata*), which is native to Central and South America but has spread throughout many tropical environments.

Artificial Cloning of Organisms

Artificial cloning of organisms may also be called *reproductive cloning*.

Methods

Reproductive cloning generally uses "somatic cell nuclear transfer" (SCNT) to create animals that are genetically identical. This process entails the transfer of a nucleus from a donor adult cell (somatic cell) to an egg that has no nucleus. If the egg begins to divide normally it is transferred into the uterus of the surrogate mother. Such clones are not strictly identical since the somatic cells may contain mutations in their nuclear DNA.

Additionally, the mitochondria in the cytoplasm also contains DNA and during SCNT this DNA is wholly from the donor egg, thus the mitochondrial genome is not the same as that of the nucleus donor cell from which it was produced. This may have important implications for cross-species nuclear transfer in which nuclear-mitochondrial incompatibilities may lead to death.

Artificial *embryo splitting* or *embryo twinning* may also be used as a method of cloning, where an embryo is split in the maturation before embryo transfer. It is optimally performed at the 6- to 8-cell stage, where it can be used as an expansion of IVF to increase the number of available embryos. If both embryos are successful, it gives rise to monozygotic (identical) twins.

Dolly the Sheep

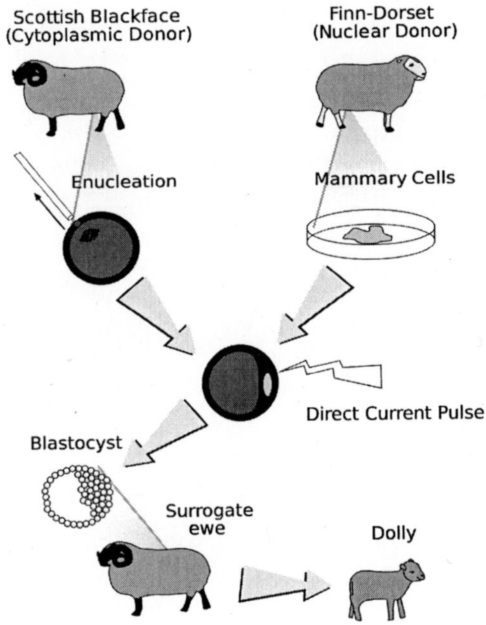

Dolly, a Finn-Dorset ewe, was the first mammal to have been successfully cloned from an adult cell. Dolly was formed by taking a cell from the udder of her biological mother. Her embryo was created by taking the cell and inserting it into a sheep ovum. The embryo was then placed inside a female sheep that went through a normal pregnancy 47. She was cloned at the Roslin Institute in Scotland and lived there from her birth in 1996 until her death in 2003 when she was six. Her stuffed remains were placed at Edinburgh's Royal Museum, part of the National Museums of Scotland.

Dolly was publicly significant because the effort showed that the genetic material from a specific adult cell, programmed to express only a distinct subset of its genes, can be reprogrammed to grow an entirely new organism. Before this demonstration, it had been shown by John Gurdon that nuclei from differentiated cells could give rise to an entire organism after transplantation into an enucleated egg.

However, this concept was not yet demonstrated in a mamallian system. Cloning Dolly the sheep had a low success rate per fertilized egg; she was born after 237 eggs were used to create 29 embryos, which only produced three lambs at birth, only one of which lived. Seventy calves have been created and one third of them died young; Prometea took 277 attempts. Notably, although the first clones were frogs, no adult cloned frog has yet been produced from a somatic adult nucleus donor cell.

There were early claims that Dolly the Sheep had pathologies resembling accelerated aging. Scientists speculated that Dolly's death in 2003 was related to the shortening of telomeres, DNA-protein complexes that protect the end of linear chromosomes. However, other researchers, including Ian Wilmut who led the team that successfully cloned Dolly, argue that Dolly's early death due to respiratory infection was unrelated to deficiencies with the cloning process.

Many people knew Dolly as a failed scientific experiment where scientists tried to control the balance of nature. Her death created awareness and allowed many people to question the ethical limits regarding cloning 47. People are mostly concerned about the risks that cloned foods may have on human health and the environment (Jesper Lassen, 2005). They fear that the applications used to create Dolly will someday be used to clone humans 47.

Water Buffalo

On September 15, 2007, the Philippines announced its development of Southeast Asia's first cloned water buffalo. The Philippine Council for Agriculture, Forestry and Natural

Resources Research and Development (PCARRD), under the Department of Science and Technology in Los Baños, Laguna approved this project.

Species Cloned

The modern cloning techniques involving nuclear transfer have been successfully performed on several species. Landmark experiments in chronological order:

- Tadpole: (1952) Many scientists questioned whether cloning had actually occurred and unpublished experiments by other labs were not able to reproduce the reported results.
- Carp: (1963) In China, embryologist Tong Dizhou produced the world's first cloned fish by inserting the DNA from a cell of a male carp into an egg from a female carp. He published the findings in a Chinese science journal.
- Mice: (1986) A mouse was the first mammal successfully cloned from an early embryonic cell. Soviet scientists Chaylakhyan, Veprencev, Sviridova, and Nikitin had the mouse "Masha" cloned. Research was published in the magazine "Biofizika" volume ÕÕÕII, issue 5 of 1987.
- Sheep: (1996) From early embryonic cells by Steen Willadsen. Megan and Morag cloned from differentiated embryonic cells in June 1995 and Dolly the sheep from a somatic cell in 1997.
- Rhesus Monkey: Tetra (January 2000) from embryo splitting
- Gaur: (2001) was the first endangered species cloned.
- Cattle: Alpha and Beta (males, 2001) and (2005) Brazil
- Cat: CopyCat "CC" (female, late 2001), Little Nicky, 2004, was the first cat cloned for commercial reasons
- Dog: Snuppy, a male Afghan hound was the first cloned dog (2005).
- Rat: Ralph, the first cloned rat (2003)
- Mule: Idaho Gem, a john mule born 4 May 2003, was the first horse-family clone.

- Horse: Prometea, a Haflinger female born 28 May 2003, was the first horse clone.
- Water Buffalo: Samrupa was the first cloned water buffalo. It was born on February 6, 2009, at India's Karnal National Diary Research Institute but died five days later due to lung infection.
- Pyrenean Ibex (2009) was the first extinct animal (extinct 2000) to be cloned back to life; the clone lived for seven minutes before dying of lung defects.
- Camel: (2009) Injaz, is the first cloned camel.

Human Cloning

Human cloning is the creation of a genetically identical copy of an existing or previously existing human. The term is generally used to refer to *artificial* human cloning; human clones in the form of identical twins are commonplace, with their cloning occurring during the natural process of reproduction. There are two commonly discussed types of human cloning: therapeutic cloning and reproductive cloning. Therapeutic cloning involves cloning adult cells for use in medicine and is an active area of research. Reproductive cloning would involve making cloned humans. A third type of cloning called replacement cloning is a theoretical possibility, and would be a combination of therapeutic and reproductive cloning. Replacement cloning would entail the replacement of an extensively damaged, failed, or failing body through cloning followed by whole or partial brain transplant.

The various forms of human cloning are controversial. There have been numerous demands for all progress in the human cloning field to be halted. Most scientific, governmental and religious organisations oppose reproductive cloning.

The American Association for the Advancement of Science (AAAS) and other scientific organisations have made public statements suggesting that human reproductive cloning be banned until safety issues are resolved. Serious ethical concerns have been raised by the future possibility of harvesting organs from clones 50. Some people have considered the idea of growing organs separately from a human organism - in doing this, a new organ supply could be established without the moral implications of harvesting them from humans. Research is also being done on the idea of growing organs that are biologically acceptable to the human body inside of other organisms, such as pigs or cows, then transplanting them to humans, a form of xenotransplantation.

The first hybrid human clone was created in November 1998, by Advanced Cell Technologies. It was created from a man's leg cell, and a cow's egg whose DNA was removed. It was destroyed after 12 days. Since a normal embryo implants at 14 days, Dr Robert Lanza, ACT's director of tissue engineering, told the Daily Mail newspaper that the embryo could not be seen as a person before 14 days. While making an embryo, which may have resulted in a complete human had it been allowed to come to term, according to ACT: "[ACT's] aim was 'therapeutic cloning' not 'reproductive cloning'"

On January, 2008, Wood and Andrew French, Stemagen's chief scientific officer in California, announced that they successfully created the first 5 mature human embryos using DNA from adult skin cells, aiming to provide a source of viable embryonic stem cells. Dr. Samuel Wood and a colleague donated skin cells, and DNA from those cells was transferred to human eggs. It is not clear if the embryos produced would have been capable of further development, but Dr. Wood stated that if that were possible, using the technology for reproductive cloning would be both unethical and illegal. The 5 cloned embryos, created in Stemagen Corporation lab, in La Jolla, were destroyed.

Ethical Issues of Cloning

Because of recent technological advancements, the cloning of animals (and potentially humans) has been an issue. The Catholic Church and many religious organisations oppose all forms of cloning, on the grounds that life begins at conception. Judaism does not equate life with conception and, though some question the wisdom of cloning, Orthodox rabbis generally find no firm reason in Jewish law and ethics to object to cloning. From the standpoint of classical liberalism, concerns also exist regarding the protection of the identity of the individual and the right to protect one's genetic identity. Gregory Stock is a scientist and outspoken critic against restrictions on cloning research. Bioethicist Gregory Pence also attacks the idea of criminalizing attempts to clone humans. The social implications of an artificial human production scheme were famously explored in Aldous Huxley's novel *Brave New World*. On December 28, 2006, the U.S. Food and Drug Administration (FDA) approved the consumption of meat and other products from cloned animals. Cloned-animal products were said to be virtually indistinguishable from the non-cloned animals.

Furthermore, companies would not be required to provide labels informing the consumer that the meat comes from a cloned animal. Critics have raised objections to the FDA's approval of cloned-animal products for human consumption, arguing that the FDA's research was inadequate, inappropriately limited, and of questionable scientific validity. Several consumer-advocate groups are working to encourage a tracking program that would allow consumers to become more aware of cloned-animal products within their food.

Joseph Mendelson, legal director of the Centre for Food Safety, said that cloned food still should be labelled since safety and ethical issues about it remain questionable. Carol Tucker Foreman, director of food policy at the Consumer Federation of America, stated that FDA does not consider the fact that the results of some studies revealed that cloned animals have increased rates of mortality and deformity at birth. A major ethical concern people have is the simple fact that manipulating animals means altering the God-given nature of things. Another concern is that the biotechnologies used on animals may someday be used on humans 47. Some people may be more open to the idea of cloning of animals because most western countries have passed legislation against cloning humans, yet only a few countries passed legislation against cloning animals.

Possible Abnormalities Due to Cloning

Researchers have found several abnormalities in cloned organisms, particularly in mice. The cloned organism may be born normal and resemble its non-cloned counterpart, but majority of the time will express changes in its genome later on in life 51. The concern with cloning humans is that the changes in genomes may not only result in changes in appearance, but in psychological and personality changes as well. The theory behind this is that the biological blueprint of the genes is the same in cloned animals as it is in normal ones, but they are read and expressed incorrectly 51. DNA arrays were used to prove this claim in the research lab of Professor Rudolf Jaenisch. Jaenisch studied placentas from cloned mice and found that one in every 25 genes was expressed abnormally 52. Results of these abnormally expressed genes in the cloned mice were premature death, pneumonia, liver failure and obesity 52.

Cloning Extinct and Endangered Species

Cloning, or more precisely, the reconstruction of functional DNA from extinct species has, for decades, been a dream of some scientists. The possible implications of this were dramatized in the best-selling novel by Michael Crichton and high budget Hollywood thriller *Jurassic Park*. In real life, one of the most anticipated targets for cloning was once the Woolly Mammoth, but attempts to extract DNA from frozen mammoths have been unsuccessful, though a joint Russo-Japanese team is currently working toward this goal.

And in January 2011, it was reported by Yomiuri Shimbun that a team of scientists headed by Akira Iritani of Kyoto University had built upon research by Dr. Wakayama, saying that they will extract DNA from a mammoth carcass that had been preserved in a Russian laboratory and insert it into the egg cells of an African elephant in hopes of producing a mammoth embryo.

The researchers said they hoped to produce a baby mammoth within six years. In 2001, a cow named Bessie gave birth to a cloned Asian gaur, an endangered species, but the calf died after two days. In 2003, a banteng was successfully cloned, followed by three African wildcats from a thawed frozen embryo. These successes provided hope that similar techniques (using surrogate mothers of another species) might be used to clone extinct species. Anticipating this possibility, tissue samples from the last *bucardo* (Pyrenean Ibex) were frozen in liquid nitrogen immediately after it died in 2000. Researchers are also considering cloning endangered species such as the giant panda, ocelot, and cheetah. The "Frozen Zoo" at the San Diego Zoo now stores frozen tissue from the world's rarest and most endangered species. In 2002, geneticists at the Australian Museum announced that they had replicated DNA of the Thylacine (Tasmanian Tiger), extinct about 65 years previous, using polymerase chain reaction. However, on February 15, 2005 the museum announced that it was stopping the project after tests showed the specimens' DNA had been too badly degraded by the (ethanol) preservative. On 15 May 2005 it was announced that the Thylacine project would be revived, with new participation from researchers in New South Wales and Victoria.

In January 2009, for the first time, an extinct animal, the Pyrenean ibex mentioned above was cloned, at the Centre of Food Technology and Research of Aragon, using the preserved DNA of the skin samples from 2001 and domestic goat egg-cells. (The ibex died shortly after birth due to physical defects in its lungs.) One of the continuing obstacles in the attempt to clone extinct species is the need for nearly perfect DNA. Cloning from a single specimen could not create a viable breeding population in sexually reproducing animals. Furthermore, even if males and females were to be cloned, the question would remain open whether they would be viable at all in the absence of parents that could teach or show them their natural behaviour.

Cloning endangered species is a highly ideological issue. Many conservation biologists and environmentalists vehemently oppose cloning endangered species—mainly because they think it may deter donations to help preserve natural habitat and wild animal populations. The "rule-of-thumb" in animal conservation is that, if it is still feasible to conserve habitat and viable wild populations, breeding in captivity should not be undertaken in isolation.

In a 2006 review, David Ehrenfeld concluded that cloning in animal conservation is an experimental technology that, at its state in 2006, could not be expected to work except by pure chance and utterly failed a cost-benefit analysis. Furthermore, he said, it is likely to siphon funds from established and working projects and does not address any of the issues underlying animal extinction (such as habitat destruction, hunting or other overexploitation, and an impoverished gene pool). While cloning technologies are well-established and used on a regular basis in plant conservation, care must be taken to ensure genetic diversity. He concluded: Vertebrate cloning poses little risk to the environment, but it can consume scarce conservation resources, and its chances of success in preserving species seem poor. To date, the conservation benefits of transgenics and vertebrate cloning remain entirely theoretical, but many of the risks are known and documented. Conservation biologists should devote their research and energies to the established methods of conservation, none of which require transgenics or vertebrate cloning.

Blastocyst

A blastocyst is formed in the early stage of the development of an embryo. During cloning process, the blastocyst cells are obtained by scientists 5 days after the egg has been divided.

In Science Fiction

Cloning has been used in countless science fiction works throughout the years. Human cloning is usually the most popular kind that is used in these particular works, mainly due to fact that it brings up controversial questions of identity. For example, in C.J. Cherryh's novel Cyteen, Aldous Huxley's novel Brave New World, human cloning is a major plot device that not only drives the story along but also makes the reader think critically about what identity means.

Star Wars also portrays human cloning in *Star Wars Episode II: Attack of the Clones* and *Star Wars Episode III: Revenge of the Sith*, in the form of the Grand Army of the Republic, an army of clone troopers. The Expanded Universe also has numerous examples of cloning, including the Thrawn trilogy, The Hand of Thrawn duology, and Clone Wars-era media.

Eukaryotic Genome

We use the term genome to refer to all of the alleles possessed by an organism (or by a population, species, or larger taxonomic group). While the amount of DNA for a diploid cell is constant within a species, the differences can be great between species. Humans have 3.5×10^9 base pairs, *Drosophila* has 1.5×10^8, toads have 3.32×10^9, and salamanders have 8×10^{10} base pairs per haploid genome. Much of the DNA in each cell either has no function or has a function not yet known.

Eukaryotes have only 10 per cent of their DNA coding for proteins. Humans may have a little as 1 per cent coding for proteins. Viruses and prokaryotes use a great deal more of their DNA. Almost half the DNA in eukaryotic cells is repeated nucleotide sequences. Protein-coding sequences are interrupted by non-coding regions. Non-coding interruptions are known as intervening sequences or introns. Coding sequences that are expressed are exons.

Most, but not all structural eukaryote genes contain introns. Although transcribed, these introns are excised (cut out) before translation (a seemingly energy inefficient process). The number of introns varies with the particular gene, even occurring in tRNAs, rRNAs and viral genes! Generally the more complex and recently evolved the organism, the more numerous and larger the introns.

Which came first: continuous genes lacking introns or interrupted genes containing introns? Introns have been hypothesized to promote genetic recombination (via crossing-over), thus speeding up the evolution of new proteins. Exons are also thought to code for different functional regions of proteins.

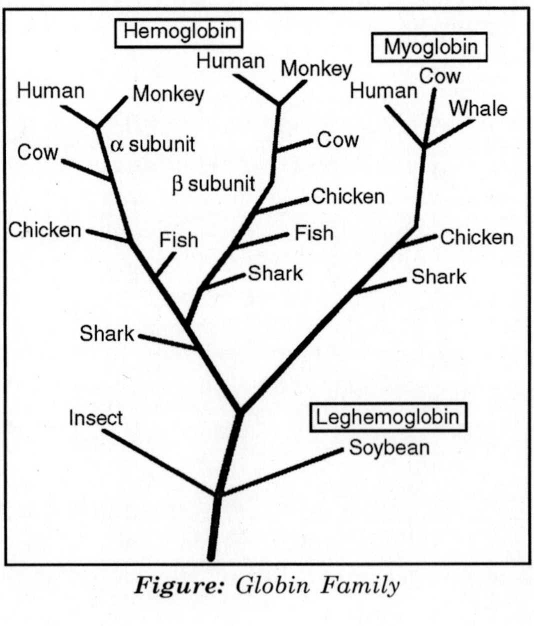

Figure: *Globin Family*

Gene families are made up of similar, but not identical, genes. The globin family is the best studied gene family. Hemoglobin consists, in humans, of 2 a-chains and 2 b-chains clustered about a common heme. Human beta-globin genes are scattered at five loci on human chromosome 11.

These genes are expressed sequentially during development, and are similar with same-length introns in similar positions in each gene. Some of the genes are inactivated copies, others are functional only during certain phases of development. The actin family of genes also exhibits a similar pattern.

Transcription and Processing of mRNA

The process of transcription in eukaryotes is similar to that in prokaryotes, although there are some differences. Eukaryote genes are not grouped in operons as are prokaryote genes. Each eukaryote gene is transcribed separately, with separate transcriptional controls on each gene. Whereas prokaryotes have one type of RNA polymerase for all types of RNA, eukaryotes have a separate RNA polymerase for each type of RNA. One enzyme for mRNA-coding genes such as structural proteins. One enzyme for large rRNAs. A third enzyme for smaller rRNAs and tRNAs.

Prokaryote translation begins even before transcription has finished, while eukaryotes have the two processes separated in time and location (remember the nuclear envelope). After eukaryotes transcribe an RNA, the RNA transcript is extensively modified before export to the cytoplasm. A cap of 7-methylguanine (a series of an unusual base) is added to the 5' end of the mRNA; this cap is essential for binding the mRNA to the ribosome. A string of adenines (as many as 200 nucleotides known as poly-A) is added to the 3' end of the mRNA after transcription.

The function of a poly-A tail is not known, but it can be used to capture mRNAs for study. Introns are cut out of the message and the exons are spliced together before the mRNA leaves the nucleus. There are several examples of identical messages being processed by different methods, often turning introns into exons and vice-versa. Protein molecules are attached to mRNAs that are exported, forming ribonucleo-protein particles (mRNPs) which may help in transport through the nuclear pores and also in attaching to ribosomes.

Antibody-coding Genes

Antibodies are globular complex proteins made by multicellular individuals in response to a specific antigen. The cells making antibodies are lymphocytes, better known as white blood cells. Antibodies immobilize and destroy their specific antigens. A mouse can make 10,000,000 antibodies, more than the total genes of the mouse would suggest. Antibody proteins are composed of two long and two short chains. Each species has a constant region characteristic for the species and type of antibody. The other region is the variable in which antigen-specific amino-acids are located.

Susumu Tonegawa tested (and proved) an old hypothesis that the constant and variable regions of antibodies were coded for by different genes. Tonegawa's work further demonstrated that gene fragments in embryos are rearranged to form the variable (functional) genes.

Viruses and Eukaryotes

The viruses of eukaryotes are similar to prokaryote-infecting viruses. Proviruses are viral DNA integrated into the host cell. Some of the DNA viruses can either initiate an infection (lytic in prokaryotes) cycle or can form proviruses. Simian Virus 40 (SV40) causes cancer in hamsters but not in its normal hosts. SV40 can introduce new functional genes into the host DNA, as can a number of other viruses.

Retroviruses can also insert their nucleic acid into host DNA via the reverse transcriptase mode mentioned previously. Reverse transcriptase is carried with the RNA into the host cell. In the process of making the cDNA strand for the viral RNA, the enzyme also makes long terminal repeats (LTRs) sequences at the terminal ends of the cDNA. These LTRs may also make insertion of the viral DNA into the host DNA easier. The inserted viral DNA makes RNA transcripts which are packaged with viral protein coats and reverse transcriptase.

The viral DNA may, depending on its insertion point, cause mutations of the host DNA. Most viral DNA insertions do not damage the host, but rather become part of the host genome and can be passed on if they have managed to infect a germ-line cell. From 0.5 to 1.0 per cent of mouse DNA may be of viral origin.

Eukaryotic Transposons

Eukaryotic transposons resemble prokaryotic transposons in many features. Many eukaryotic transposons are first copied to RNA and then back to DNA before being inserted in a new location.

Genes, Viruses and Cancer

Cancer is a disease in which cells escape the restraints on normal cell growth. Cancer is an inheritable disease (at least from cell to daughter cells). Once a cell has become cancerous, all of its descendant cells are cancerous. Gross chromosomal abnormalities are often visible in cancerous cells. Most carcinogens (cancer-generating factors) are also mutagens (mutation-generating factors). Oncogenes are genes resembling normal genes but in which something has gone wrong, resulting in a cancer. Fifty oncogenes have thus far been discovered. Viruses seem able to cause cancer in three ways.

Presence of the viral DNA may disrupt normal host DNA functions. Viral proteins needed for virus replication may also affect normal host gene regulation.

Since most cancer-causing viruses are retroviruses, the virus may serve as a vector for oncogene insertion. Transfers of genes between eukaryotic cells will allow doctors, who have historically been limited to phenotypic cures, to attack disease at the genotypic level. SV40 virus has been used to inject the rabbit beta-globin gene into monkeys. Viruses can thus serve as a possible vector to place healthy (non-mutated) alleles into eggs.

Like prokaryotes, eukaryotic organisms do not want to express all of their genes all of the time. Given the complexity of multicellular eukaryotes, gene regulation in these organisms needs to be very complex. This module provides a brief overview of the various levels of regulation of gene expression in eukaryotes, and takes a look at the basics of transcriptional regulation; a detailed look at eukaryotic transcriptional regulation is beyond the scope of this course.

Gene Regulation in Eukaryotes

Eukaryotes need to regulate their genes for different reasons than prokaryotes. In prokaryotes, gene regulation allowed them to respond to their environment efficiently and economically. While eukaryotes can respond to their environment, the main reason higher eukaryotes need to regulate their genes is cell specialization.

Whereas prokaryotes are (relatively speaking) simple, unicellular organisms, multicellular eukaryotes consist of hundreds of different cell types, each differentiated to serve a different specialized function. Each cell type differentiates by activating a different subset of genes. Because of the multitude of cell types, the regulation of gene expressionrequired to bring about such differentiation is necessarily complex. One way this complexity is demonstrated is in multiple levels of regulation of gene expression.

Levels of Regulation

Before we discuss the specifics of regulation, it is necessary to understand that "gene expression" covers the entire process from transcription through protein synthesis. The final measure of whether or not a gene is "expressed" is if the protein is produced, because it is protein that will ultimately carry out the function specified by the gene.

We've seen numerous examples of how eukaryotic cells are more complex than prokaryotic cells. One obvious example of this is the presence of a nucleus in eukaryotic cells, which separates transcription from translation in a way not seen in prokaryotes. Furthermore, eukaryotic transcripts must be processed before they can be translated. Here is a diagram outlining the steps involved in the production of a protein in eukaryotic cells: Regulation can occur at any point in this pathway; specifically, it occurs at the levels of transcription, RNA processing, mRNA lifetime (longevity), and translation. Each of these types of regulation will be considered in turn.

Regulation of RNA Processing

After transcription, the RNA must be processed before it can be translated. As described elsewhere, RNA processing involves addition of a 5' cap, addition of a 3' poly (A) tail, and removal of introns. This processing represents another level of regulation of gene expression, particularly in regard to splicing out of introns.

Regulation can be of two types:

1. Whether an RNA gets processed;
2. Which exons are retained in the mRNA.

The first type of regulation can determine whether or not an mRNA gets translated. If an RNA is not processed, it will not be transported out of the nucleus, and will not be translated. The second type of regulation can affect the function of the protein produced. Some genes have exons that can be exchanged in a process known as exon shuffling. For example, a gene with four exons might be spliced differently in two different cell types.

In each of these cases, the polypeptide produced could have a different function. In mammals, for example, the calcitonin gene produces a hormone in one cell type, and a neurotransmitter in another cell type, due to exon shuffling.

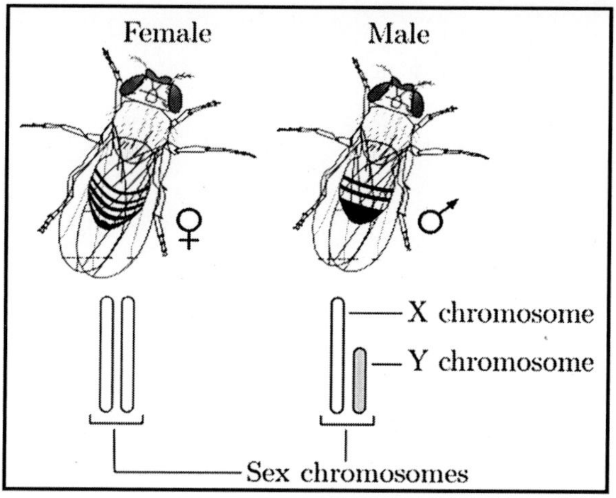

Figure: *Drosophila*

In *Drosophila*, alternate splicing of the *sex-lethal* RNA can produce an mRNA encoding a functional polypeptide, or one with a premature stop codon that encodes a short, nonfunctional polypeptide.

Regulation of RNA Longevity

Imagine two mRNA molecules: one lasts for five minutes in the cytoplasm before being degraded, while the other one manages to linger for an hour before being degraded. If both are translated continually while they exist, it is obvious that more of the second polypeptide will be produced than the first. This is the principle behind regulation of RNA longevity. mRNAs from different genes have their approximate lifespan encoded in them; this serves to help regulate how much of each polypeptide is produced.

The information for lifespan is found in the 3' UTR. The sequence AUUUA, when found in the 3' UTR, is a signal for early degradation (and therefore short lifetime). The more times the sequence is present, the shorter the lifespan of the mRNA. Because it is encoded in the nucleotide sequence, this is a set property of each different mRNA; the longevity of an mRNA can't be varied.

Chapter 3

Nano Stem Cell Technology

Genetic Disorders

A genetic disorder is an illness caused by abnormalities in genes or chromosomes, especially a condition that is present from before birth. Most genetic disorders are quite rare and affect one person in every several thousands or millions.

A genetic disorder may or may not be a heritable disorder. Some genetic disorders are passed down from the parents' genes, but others are always or almost always caused by new mutations or changes to the DNA.

In other cases, the same disease, such as some forms of cancer, may be caused by an inherited genetic condition in some people, by new mutations in other people, and by non-genetic causes in still other people. Some types of recessive gene disorders confer an advantage in the heterozygous state in certain environments.

Single Gene Disorder

A single gene disorder is the result of a single mutated gene. There are estimated to be over 4000 human diseases caused by single gene defects. Single gene disorders can be passed on to subsequent generations in several ways. Genomic imprinting and uniparental disomy, however, may affect inheritance patterns. The divisions between recessive and dominant types are not "hard and fast" although the divisions between autosomal and X-linked types are (since the latter types are distinguished purely based on the chromosomal location of the gene).

For example, achondroplasia is typically considered a dominant disorder, but children with two genes for achondroplasia have a severe skeletal disorder that achondroplasics could be viewed as carriers of. Sickle-cell anemia is also considered a recessive condition, but heterozygous carriers have increased resistance to malaria in early childhood, which could be described as a related dominant condition. When a couple where one partner or both are sufferers or carriers of a single gene disorder and wish to have a child they can do so through IVF which means they can then have PGD (pre-implantation genetic diagnosis) to check whether the fertilized egg has had the genetic disorder passed on.

Autosomal Dominant

Only one mutated copy of the gene will be necessary for a person to be affected by an autosomal dominant disorder. Each affected person usually has one affected parent. There is a 50% chance that a child will inherit the mutated gene. Conditions that are autosomal dominant sometimes have reduced penetrance, which means that although only one mutated copy is needed, not all individuals who inherit that mutation go on to develop the disease.

Examples of this type of disorder are Huntington's disease, neurofibromatosis type 1, neurofibromatosis type 2, Marfan syndrome, hereditary nonpolyposis colourectal cancer, and hereditary multiple exostoses, which is a highly penetrant autosomal dominant disorder. Birth defects are also called congenital anomalies.

Dominance (Genetics)

Dominance in genetics is a relationship between two variant forms (alleles) of a single gene, in which one allele masks the effect of the other in influencing some trait. In the simplest case, if a gene exists in two allelic forms (*A* & *a*), three combinations of alleles (genotypes) are possible: *AA*, *Aa*, and *aa*. If *Aa* individuals (heterozygotes) show the same form of the trait (phenotype) as *AA* individuals (homozygotes), and *aa* homozygotes show a different phenotype, allele *A* is said to *dominate* or *be dominant to* allele *a*, and *a* is said to *be recessive to A*.

Background: Diploid, Chromosomes, Genes, Loci, & Alleles

Diploid / Haploid

Most familiar plants, like peas, and familiar animals, like fruit flies and humans, have paired chromosomes, and are described as diploid. One chromosome of each pair is contributed by each parent: one by the female parent in her ovum, and one by the male parent in his sperm, which are joined at fertilization. The ovum and sperm cells have only one copy of each chromosome and are described as haploid. Production of haploid gametes occurs through a process called meiosis.

Chromosomes, Genes, and Alleles

Each chromosome of a matching pair is structurally similar to the other, and each member of a homologous pair has the same genetic material arranged in the same order and physical locations (loci, singular locus). The genetic material in each chromosome comprises a series of discrete genes that influence various traits.

Thus, each gene also has a corresponding homologue, which may exist in different forms: the variant forms are called alleles. The alleles at the same locus on the two homologous chromosomes may be identical or different.

In popular usage, "gene" and "allele" are often used interchangeably. This produces misunderstandings. Properly, "gene" refers to a hereditary unit, ordinarily at a fixed position on a chromosome, that influences a particular trait. Genes are now understood to comprise DNA. "Allele" refers to any of the many particular forms of a gene that may be present in a population of individuals from a particular species, at a particular locus.

E.g., it is inaccurate to say "This pea plant has a pair of wrinkled genes", and it is more accurate to say, "This plant has two 'w' alleles for the 'Seed Shape' gene, and will produce wrinkled peas." Consider also the example of blood type in humans. Near the long arm of chromosome nine appears a gene that determines whether an individual will be blood type, A, B, or O. There are three different alleles that could be present at this locus, but only two can be present in any individual, one inherited from their mother and one from their father.

Homozygous, Heterozygous

If two alleles of a given gene are identical, the organism is called a homozygote and is homozygous with respect to that gene; if instead the two alleles are different, the organism is a heterozygote and is heterozygous. The genetic makeup of an organism, either at a single locus or over all its genes collectively, is called the genotype.

The genotype of an organism directly or indirectly affects its molecular, physical and other traits, which individually or collectively are called the phenotype. At heterozygous gene loci, the two alleles interact to produce the phenotype. The simplest form of allele interaction is the one described by Mendel, now called Mendelian, in which the appearance/phenotype caused by one allele is apparent, called dominant, and the appearance/ phenotype caused by the other allele is not apparent, called recessive.

In the simplest case, the phenotypic effect of one allele completely masks the other in heterozygous combination; that is, the phenotype produced by the two alleles in heterozygous combination is identical to that produced by one of the two homozygous genotypes. The allele that masks the other is said to be *dominant* to the latter, and the alternative allele is said to be *recessive* to the former.

Which Trait is **Dominant?**

The terms *dominant* and *recessive* refer to the interaction of alleles in producing the phenotype of the heterozygote. If there are two alternative phenotypes, by definition the phenotype exhibited by the heterozygote is called "dominant" and the "hidden" phenotype is called "recessive". The key concept of dominance is that the heterozygote is phenotypically *identical* to one of the two homozygotes.

That trait corresponding to the dominant allele may then be called the "dominant" trait. Dominance is a *genotypic* relationship between *alleles*, as manifested in the phenotype. It is unrelated to the nature of the phenotype itself, e.g., whether it is regarded as normal or abnormal, standard or nonstandard, healthy or diseased, stronger or weaker, or more or less extreme. It is also important to distinguish between the "round" gene locus, the "round" allele at that locus, and the "round" phenotype it produces. It is inaccurate to say that "the round gene dominates the wrinkled gene" or that "round peas dominate wrinkled peas."

Nomenclature

In genetics, the common convention is that dominant alleles are written as capital letters and recessive alleles as lower-case letters. In the pea example, once the dominance relationships of the two alleles are known, it is possible to designate the dominant allele that produces a round shape by a capital-letter symbol R, and the alternative recessive allele that produces a wrinkled shape by a lower-case symbol r. The homozygous dominant, heterozygous, and homozygous recessive genotypes are then written RR, Rr, and rr, respectively. It would also be possible to designate the two alleles as W and w, and the three genotypes WW, Ww, and ww, the first two of which produced round peas and the third wrinkled peas. Note that the choice or "R" or "W" as the symbol for the dominant allele does not prejudge whether the allele causing the "round" or "wrinkled" phenotype when homozygous is the dominant one.

Another system of notation designates the gene involved in seed shape as the "*Shp*" gene, which exists in two allelic forms, Shp^R and Shp^w, the dominance relationships of the two being indicated by the case of the superscripts. This system is the standard system in *Drosophila* genetics.

Relationship to other Genetic Concepts

The concept of dominance is involved with a number of other genetic concepts.

Multiple Alleles

Although any individual has at most two different alleles, most genes exist in a large number of allelic forms in the population as a whole. In some cases, the alleles have different effects on the phenotype, and their dominance interactions with each other can be described as a series. For example, the best known human blood groups, the ABO system, comprises three sets of alleles at the *I* locus, I^A, I^B, and I^O. The first two are dominant to the latter:

that is, the AA and AO genotypes produce indistinguishable blood group phenotypes, called "*Type A*", as do BB and BO, which produce "*Type B*" blood. In another example, coat colour in siamese cats and related breeds is determined by a series of alleles at the albino gene locus (*c*) that produce different levels of pigment and hence different levels of colour dilution. Four of these are c^+, c^b, c^s, and c^a (standard, Burmese, siamese, and albino, respectively), where the first allele is completely dominant to the last three, and the last is completely recessive to the first three.

Incomplete and Semi-dominance

Complete dominance occurs when the phenotype of the heterozygote is completely indistinguishable from that of the dominant homozygote. This is frequently not the case. Incomplete dominance occurs when the phenotype of the heterozygous genotype is an intermediate of the phenotypes of the homozygous genotypes. For example, the snapdragon flower colour is either homozygous for red or white. When the red homozygous flower is paired with the white homozygous flower, the result yields a pink snapdragon flower. The pink snapdragon is the result of incomplete dominance.

Co-dominance

Co-dominance occurs when the contributions of both alleles are visible in the phenotype. In the ABO example, the I^A and I^B alleles are co-dominant in producing the AB blood group phenotype, in which both A-type and B-type antigens are made. Another example occurs at the locus for the Beta-globin component of hemoglobin, where the three molecular phenotypes of Hb^A/Hb^A, Hb^A/Hb^S, and Hb^S/Hb^S are all equally detectable by protein electrophoresis. (The medical condition produced by the heterozygous genotype is called an *incomplete dominant*). For most gene loci at the molecular level, both alleles are expressed co-dominantly, because both are transcribed into RNA.

Co-dominance is different from incomplete or semi-dominance. For example, pink flowers might be the product of two alleles that produce red and white pigments that become mixed (co-dominance on the pigment level, no dominance on the colour level), or the result of one allele that produces the usual amount of red pigment and another non-functional allele that produces no pigment, so as to produce a dilute, intermediate pink colour (no dominance at either level).

Autosomal Versus Sex-linked Dominance

In humans and other mammal species, sex is determined by two sex chromosomes called the X chromosome and the Y chromosome. Human females are typically XX; males are typically XY. The remaining pairs of chromosome are found in both sexes and are called autosomes; genetic traits due to loci on these chromosomes are described as autosomal, and may be dominant or recessive. Genetic traits on the X and Y chromosomes are called sex linked, because they tend to be characteristic of one sex or the other.

In practice, the term almost always refers to X-linked traits. Females have two copies of every gene locus found on the X chromosome, just as for the autosomes, and the same dominance relationships apply. Males however have only one copy of each X chromosome gene locus, and are described as hemizygous for these genes. The Y chromosome is much smaller than the X, and contains a much smaller set of genes that influence 'maleness', such as the SRY gene for testis determining factor. Dominance rules for sex-linked gene loci are determined by their behaviour in the female: because the male has only one allele, that allele is always expressed regardless of whether it is dominant or recessive. An example of sex-linked is: Hair loss is a trait caused by a gene located on the X chromosomes.

Epistasis

Epistasis ["*epi* + *stasis* = to sit on top"] is an interaction between genotypes at two *different* gene loci, which sometimes resembles a dominance interaction at a single locus. Epistasis modifies the characteristic 9:3:3:1 ratio expected for two non-epistatic genes. Most genetic systems involve complex epistatic interactions among multiple gene loci. For two loci, 14 classes of epistatic interactions are recognised. As an example of *recessive epistasis*, one gene locus may determine whether a flower pigment is yellow (AA or Aa) or green (aa), while another locus determines whether the pigment is produced (BB or Bb) or not (bb). In a bb plant, the flowers will be white, irrespective of the genotype of the other locus as AA, Aa, or aa. The b allele is *not* dominant to the A allele: the B locus shows *recessive epistasis* to the A locus, because the B locus when homozygous for the recessive allele (bb) suppresses phenotypic expression of the A locus. In a cross between two AaBb plants, this produces a characteristic 9:3:4 ratio, in this case of yellow : green : white flowers. In *dominant epistasis*, one gene locus may determine yellow and green pigment as in the previous example: AA and Aa are yellow, and aa are green.

A second locus determines whether a pigment precursor is produced (dd) or not (DD or Dd). Here, in a D- plant, the flowers will be colourless irrespective of the genotype at the *A* locus, because of the epistatic effect of the dominant D allele. Thus, in a cross between two AaDd plants, 3/4 of the plants will be colourless, and the yellow and green phenotypes are expressed only in dd plants. This produces a characteristic 12:3:1 ratio of white : yellow : green plants.

Identification of Genetic Material

The genetic material of a cell or an organism refers to those materials found in the nucleus, mitochondria and cytoplasm, which play a fundamental role in determining the structure and nature of cell substances, and capable of self-propagating and variation. When it became evident that the chromosomes were the organs of heredity, because (i) they formed the only link between two generations; (ii) they carried linearly arranged genes; (iii) they occurred in every organism in specific number and had specific morphology for a particular species; and (iv)any variation in their number or morphology affected the phenotype of the species. Then, various attempts were made by early molecular geneticists to identify the physical and chemical nature of genes. But, the genes were found so minute structures that their physical identity remained almost impossible. However, the extensive chemical analysis of chromosomes of different organisms have revealed that chromosomes contain proteins and nucleic acid (DNA and RNA) and it was thought that genes might have either proteins or nucleic acids as their component molecules.

Early molecular geneticists have assigned the informational roles of genes to the chromosomal proteins, because, they found nucleic acids too simple to carry genetic informations.

The controversy about the assignment of genetic role either to chromosomal proteins or to chromosomal DNA, existed upto 1950, when finally it was universally accepted that DNA is the genetic substance (I. e., chemical of which genes are composed) of most micro-organisms and higher organisms.Later on, RNA was found to be the genetic material of some viruses. The concept that DNA or RNA is the genetic material of most organisms has been developed and supported by following evidences:

Direct Evidences for DNA as the Genetic Material

The most conclusive evidences in support of DNA as the genetic material comes from following three avenues of approach on micro-organisms: transformation of bacteria mode of infection of bacteriophages and conjugation in bacteria. Bacterial Transformation and Griffith EffectIn 1928, Frederick Griffith encountered a phenomenon now known as genetic transformation. Colonies of virulent strain (pathogenic) of pneumonia causing spherically-shaped bacterium, Diplococcus pneumoniae grown on nutrient agar, have a smooth (s) glistering appearance owing to the presence of a type specific, ploysaccharide (a polymer of glucose and glucuronic acid) capsule.

The avirulant (non-pathogenic) strains of this bacterium, on the other hand, lack this capsule and they produce dull, rough (R) colonies and is unable to cause the pneumonia disease. Smooth (S) and rough (R) characters are directly related to the presence or absence of the capsule and this trait is known to be genetically determined. Both S and R forms occur in several types and are designed as S-1, S-II, S-III, etc., and R-I, R-II, R-III, etc., respectively.

All these subtypes of S and R bacteria differ with each other in the type of antigens, they produce. The kind of antigens produced is likewise genetically determined. Smooth (S) forms sometimes mutate to rough (R) forms, but this change have not been found reversible.

In the course of his work, Griffith injected laboratory mice with live R-II pneumococci; the mice suffered no illness because R-II pneumococci was avirulent.

But when the mice were injected with the mixture of living avirulent R-II and heat killed S-III virulent, the unexpected symptoms of pneumonia appeared and high mortality was resulted in them.

By postmortoming the dead mice, it was found that their heart blood had both R-II and S-III pneumococci. From these results, Griffith concluded that the presence of the heat-killed S-IU bacteria must have caused a transformation of the living R-II bacteria, so as to restore to them the capacity for capsule formation they had earlier lost by gene mutation.

This was called "Griffith effect" or more popularly as "bacterial transformation". Identification of the "transforming substance"-Griffith could not understand the cause of bacterial transformation and that is first of all identified by O. T. Avery, C. M. McLeod and M McCarthy (1944). They tested a fraction of heat killed S-III bacteria for the transforming ability.

They could remove proteins, lipids, polysaccharides and ribonucleic acids from the S-III extract by a variety of chemical and enzymatic methods without seriously diminishing its power to transfer R-II mutants into the S-III wild types. They ultimately concluded that because a cell free and highly purified DNA extract of S-III bacteria could bring about transformation of R-I1 bacteria into S-III, therefore, DNA is the genetic material of pneumococci.

Later on, such a transforming substance, the DNA, was found in a variety of bacteria (i.e., Hemophilus influenzae. Bacillus subtilus, Shigelia paradysenteriae, etc.) and several other organisms.

Mode of Infection of Bacteriophages and Identification of DNA as their Genetic Material

By using radioactive tracers, A. Hershey and M. Chase (1952) provided further direct proof that DNA is the genetic material in certain bacterial viruses. These investigators were studying the bacteriophages that attack the bacterium, Escherichia coli. They prepared a chemically defined culture medium (for E. coli) containing phosphoric acid and sulphuric acid as the only source of phosphorus and sulphur.

Known quantities of radioactive isotopes of phosphorus (P^{32}) and sulphur (S^{32}) were added to the culture medium. E. coli grown on such a culture medium incorporated the P^{32} and S^{35} into their chemical constituents. When T_2 bacteriophage particles infected the radioactive bacteria, they incorporated the labeled DNA.Phage capsid proteins did not contain appreciable amounts of phosphorus and only the DNA was labeled with P^{32}. Similarly, the protein envelope around the phage was selectively labeled with S^{35}. DNA contains virtually no sulphur and was not labeled with S^{35} By this method it was possible to differentially label the phage DNA and proteins of phage-capsid. After phage growth, the phage particles were separated from the host cell by centrifugation. The radioactive phages were next introduced to non-radioactive bacterial cultures where they attacked the bacteria.

Subsequently the viruses were separated from the host cell by agitation and the content of P^{32} and S^{35} in the host (E. coli) and parasite (bacteriophage) was determined. The phosphorus label was found to be associated with the bacterial cells and the sulphur label was in the proteins coats (capsids) left in the medium. This indicated that the DNA had panetrated the cells but that the protein coat or capsid of the phage was left outside the wall of the bacterium. The labeled DNA was found to reproduced in the host cells. These experiments, thus, clearly demonstrated that genetic informations of these bacteriophages resided in their DNA molecules and not in the protein molecules of their capsids.

Later on Hershey-Chase experiment was repeated for other viruses and it was found that DNA is also the genetic material in a number of other viruses.

Bacterial Conjugation

Another conclusive evidence for DNA as the genetic material came from the phenomenon of conjugation of bacteria. Laderberg and Tatum (1946) found that when a F^+ ('male') E. coli cell conjugated with a F^- ('female') E. coli cell an unidirectional transfer of F^+ factor of 'male' cell to F^- or 'female' cell took place; so that the latter was converted into a F^+ or 'male' strain. The F^+ factor was found to be a fragment of DNA molecule which occurred in the cytoplasm of bactarial cell.

Genetically Modified Organism

A genetically modified organism (GMO) or genetically engineered organism (GEO) is an organism whose genetic material has been altered using genetic engineering techniques. These techniques, generally known as recombinant DNA technology, use DNA molecules from different sources, which are combined into one molecule to create a new set of genes. This DNA is then transferred into an organism, giving it modified or novel genes. Transgenic organisms, a subset of GMOs, are organisms which have inserted DNA that originated in a different species.

Production

Genetic modification involves the insertion or deletion of genes. When genes are inserted, they usually come from a different species, which is a form of horizontal gene transfer.

Figure: *GloFish, the first genetically modified animal to be sold as a pet*

In nature this can occur when exogenous DNA penetrates the cell membrane for any reason. To do this artificially may require attaching the genes to a virus or just physically inserting the extra DNA into the nucleus of the intended host with a very small syringe, or with very small particles fired from a gene gun. However, other methods exploit natural forms of gene transfer, such as the ability of Agrobacterium to transfer genetic material to plants, or the ability of lentiviruses to transfer genes to animal cells.

History

The general principle of producing a GMO is to add new genetic material into an organism's genome. This is called genetic engineering and was made possible through the discovery of DNA and the creation of the first recombinant bacteria in 1973; an existing bacterium E. coli expressing an exogenic Salmonella gene.

This led to concerns in the scientific community about potential risks from genetic engineering, which were first discussed in depth at the Asilomar Conference in 1975. One of the main recommendations from this meeting was that government oversight of recombinant DNA research should be established until the technology was deemed safe. Herbert Boyer then founded the first company to use recombinant DNA technology, Genentech, and in 1978 the company announced creation of an E. coli strain producing the human protein insulin. In 1986, field tests of bacteria genetically engineered to protect plants from frost damage (ice-minus bacteria) at a small biotechnology company called Advanced Genetic Sciences of Oakland, California, were repeatedly delayed by opponents of biotechnology. In the same year, a proposed field test of a microbe genetically engineered for a pest resistance protein by Monsanto Company was dropped.

In the late 1980s and early 1990s, guidance on assessing the safety of genetically engineered plants and food emerged from organizations including the FAO and WHO. Small scale experimental plantings of genetically modified (GM) plants began in Canada and the U.S. in the late 1980s. The first approvals for large scale, commercial cultivation came in the mid 1990s. Since that time, adoption of GM plants by farmers has increased annually.

Uses

GMOs are used in biological and medical research, production of pharmaceutical drugs, experimental medicine (e.g. gene therapy), and agriculture (e.g. golden rice). The term "genetically modified organism" does not always imply, but can include, targeted insertions of genes from one species into another. For example, a gene from a jellyfish, encoding a fluorescent protein called GFP, can be physically linked and thus co-expressed with mammalian genes to identify the location of the protein encoded by the GFP-tagged gene in the mammalian cell. Such methods are useful tools for biologists in many areas of research, including those who study the mechanisms of human and other diseases or fundamental biological processes in eukaryotic or prokaryotic cells.

To date the most controversial but also the most widely adopted application of GMO technology is patent-protected food crops which are resistant to commercial herbicides or are able to produce pesticidal proteins from within the plant, or stacked trait seeds, which do both. The largest share of the GMO crops planted globally are owned by the US firm Monsanto.

In 2007, Monsanto's trait technologies were planted on 246 million acres (1,000,000 km) throughout the world, a growth of 13 percent from 2006. However, patents on the first Monsanto products to enter the marketplace will begin to expire in 2014, democratizing Monsanto products. In addition, a 2007 report from the European Joint Research Commission predicts that by 2015, more than 40 per cent of new GM plants entering the global marketplace will have been developed in Asia.

In the corn market, Monsanto's triple-stack corn—which combines Roundup Ready 2 weed control technology with YieldGard Corn Borer and YieldGard Rootworm insect control—is the market leader in the United States. U.S. corn farmers planted more than 32 million acres (130,000 km) of triple-stack corn in 2008, and it is estimated the product could be planted on 56 million acres (230,000 km) in 2014–2015. In the cotton market, Bollgard II with Roundup Ready Flex was planted on approximately 5 million acres (20,000 km) of U.S. cotton in 2008.

According to the International Service for the Acquisition of Agri-Biotech Applications (ISAAA), in 2010 approximately 15 million farmers grew biotech crops in 29 countries. Over 90% of the farmers were resource-poor in developing countries.

6.5 million farmers in China and 6.3 million small farmers in India grew biotech crops (mostly Bacillus thuringiensis cotton). The Philippines, South Africa (biotech cotton, maize and soybeans often grown by subsistence women farmers) and another twelve developing countries also grew biotech crops in 2009. 10 million more small and resource-poor farmers may have been secondary beneficiaries of Bt cotton in China. The global commercial value of biotech crops grown in 2008 was estimated to be US$130 billion. In the United States, the United States Department of Agriculture (USDA) reports on the total area of GMO varieties planted. According to National Agricultural Statistics Service, the states published in these tables represent 81–86 percent of all corn planted area, 88–90 percent of all soybean planted area, and 81–93 percent of all upland cotton planted area (depending on the year).

USDA does not collect data for global area. Estimates are produced by the International Service for the Acquisition of Agri-biotech Applications (ISAAA) and can be found in the report, "Global Status of Commercialized Transgenic Crops: 2007". Transgenic animals are also becoming useful commercially. On February 6, 2009 the U.S. Food and Drug Administration approved the first human biological drug produced from such an animal, a goat. The drug, ATryn, is an anticoagulant which reduces the probability of blood clots during surgery or childbirth. It is extracted from the goat's milk.

Detection

Testing on GMOs in food and feed is routinely done by molecular techniques like DNA microarrays or qPCR. The test can be based on screening elements (like p35S, tNos, pat, or bar) or event-specific markers for the official GMOs (like Mon810, Bt11, or GT73). The array-based method combines multiplex PCR and array technology to screen samples for different potential GMOs, combining different approaches (screening elements, plant-specific markers, and event-specific markers).

The qPCR is used to detect specific GMO events by usage of specific primers for screening elements or event-specific markers.

To avoid any kind of false positive or false negative testing outcome, comprehensive controls for every step of the process is mandatory. A CaMV check is important to avoid false positive outcomes based on virus contamination of the sample.

Transgenic Microbes

Bacteria were the first organisms to be modified in the laboratory, due to their simple genetics. These organisms are now used for several purposes, and are particularly important in producing large amounts of pure human proteins for use in medicine.

Genetically modified bacteria are used to produce the protein insulin to treat diabetes. Similar bacteria have been used to produce clotting factors to treat haemophilia, and human growth hormone to treat various forms of dwarfism.

Transgenic Animals

Figure: *Some chimeras, like the blotched mouse shown, are created through genetic modification techniques like gene targeting.*

Transgenic animals are used as experimental models to perform phenotypic and for testing in biomedical research.

Genetically Modified (Genetically Engineered) animals are becoming more vital to the discovery and development of cures and treatments for many serious diseases. By altering the DNA or transferring DNA to an animal, we can develop certain proteins that may be used in medical treatment. Stable expressions of human proteins have been developed in many animals, including sheep, pigs, and rats.

Some examples are: Human-alpha-1-antitrypsin, which has been developed in sheep and is used in treating humans with this deficency and transgenic pigs with human-histo-compatibility have been studied in the hopes that the organs will be suitable for transplant with less chances of rejection.

Transgenic livestock have been used as bioreactors since the 1990s. Many medicines, including insulin and many immunizations are developed in transgenic animals. In March 2011, the bioactive recombinant Human Lysozyme was expressed in the milk of cloned transgenic cattle. This field is growing rapidly and new pharming uses are being discovered and developed. The extent that trangenic animals will be useful in the medical field as well as other fields is very promising based on results thus far.

Fruit Flies

In biological research, transgenic fruit flies (Drosophila melanogaster) are model organisms used to study the effects of genetic changes on development. Fruit flies are often preferred over other animals due to their short life cycle, low maintenance requirements, and relatively simple genome compared to many vertebrates.

Mosquitoes

In 2010, scientists created "malaria-resistant mosquitoes" in the laboratory. The World Health Organisation estimated that Malaria killed almost one million people in 2008. Genetically modified male mosquitoes containing a lethal gene have been developed in order to combat the spread of Dengue fever.

Aedes aegypti mosquitoes, the single most important carrier of dengue fever, were reduced by 80% in a 2010 trial of these GM mosquitoes in the Cayman Islands. Between 50 - 100 million people are affected by Dengue fever every year and 40,000 people die from it.

Bollworms

A strain of Pectinophora gossypiella (Pink bollworm) has been developed that contains a fluorescent marker in their

DNA. This allows researchers to monitor bollworms that have been sterilized by radiation and released in order to reduce bollworm infestation.

Mammals

Genetically modified mammals are an important category of genetically modified organisms. Transgenic mice are often used to study cellular and tissue-specific responses to disease.

In 1999, scientists at the University of Guelph in Ontario, Canada created the genetically engineered Enviropig.

The Enviropig excretes from 30 to 70.7% less phosphorus in manure depending upon the age and diet. In February 2010, Environment Canada determined that Enviropigs are in compliance with the Canadian Environmental Protection Act and can be produced outside of the research context in controlled facilities where they are segregated from other animals.

In 2009, scientists in Japan announced that they had successfully transferred a gene into a primate species (marmosets) and produced a stable line of breeding transgenic primates for the first time. Their first research target for these marmosets was Parkinson's disease, but they were also considering Amyotrophic lateral sclerosis and Huntington's disease.

In 2011, scientists in China released news that they have introduced human genes into 300 dairy cows to produce milk with the same properties as human breast milk. Aside from milk production, the researchers claim these transgenic cows to be identical to regular cows.

Cnidarians

Cnidarians such as Hydra and the sea anemone Nematostella vectensis have become attractive model organisms to study the evolution of immunity and certain developmental processes. An important technical breakthrough was the development of procedures for generation of stably transgenic hydras and sea anemones by embryo microinjection.

Fish

Genetically modified fish have promoters driving an over-production of "all fish" growth hormone. This resulted in dramatic growth enhancement in several species, including salmonids, carps and tilapias.

Gene Therapy

Gene therapy, uses genetically modified viruses to deliver genes that can cure disease into human cells. Although gene therapy is still relatively new, it has had some successes.

It has been used to treat genetic disorders such as severe combined immunodeficiency, and treatments are being developed for a range of other currently incurable diseases, such as cystic fibrosis, sickle cell anemia, Parkinson's disease and muscular dystrophy.

Current gene therapy technology only targets the non-reproductive cells meaning that any changes introduced by the treatment can not be transmitted to the next generation. Gene therapy targeting the reproductive cells—so-called "Germ line Gene Therapy"—is very controversial and is unlikely to be developed in the near future.

Transgenic Plants

Transgenic plants have been engineered to possess several desirable traits, such as resistance to pests, herbicides, or harsh environmental conditions, improved product shelf life, and increased nutritional value. Since the first commermmodified to be tolerant to the herbicides glufosinate and glyphosate, to be resistant to virus damage as in Ringspot virus-resistant GM papaya, grown in Hawaii, and to produce the Bt toxin, an insecticide that is non-toxic to mammals. Most GM crops grown today have been modified with "input traits", which provide benefits mainly to farmers. The GM oilseed crops on the market today offer improved oil profiles for processing or healthier edible oils.

The GM crops in development offer a wider array of environmental and consumer benefits such as nutritional enhancement, drought and stress tolerance. GM plants are being developed by both private companies and public research institutions such as CIMMYT, the International Maize and Wheat Improvement Centre. Other examples include a genetically modified sweet potato, enhanced with protein and other nutrients, while golden rice, developed by the International

Rice Research Institute (IRRI), has been discussed as a possible cure for Vitamin A deficiency.

Scientists at the University of York developed a weed (Arabidopsis thaliana) which contained genes from bacteria that can clean up TNT and RDX explosive contaminants from the soil and it was hoped that this weed would eliminate this pollution. 16 million hectares in the USA (1.5% of the total surface) are estimated to be contaminated with TNT and RDX. However the weed Arabidopsis thaliana was not tough enough to withstand the environment on military test grounds and research is continuing with the University of Washington to develop a tougher native grass.

The coexistence of GM plants with conventional and organic crops has raised significant concern in many European countries. Due to relatively high demand from European consumers for the freedom of choice between GM and non-GM foods, EU regulations require measures to avoid mixing of foods and feed produced from GM crops and conventional or organic crops.

European research programs such as Co-Extra, Transcontainer, and SIGMEA are investigating appropriate tools and rules. At the field level, biological containment methods include isolation distance and pollen barriers. Such measures are generally not used in North America because they are very costly and there are no safety-related reasons to employ them.

Cisgenic Plants

Cisgenesis, sometimes also called Intragenesis, is a product designation for a category of genetically engineered plants. A variety of classification schemes have been proposed, that order genetically modified organisms based on the nature of introduced genotypical changes rather than the process of genetic engineering.

While some genetically modified plants are developed by the introduction of a gene originating from distant, sexually incompatible species into the host genome, cisgenic plants contain genes which have been isolated either directly from the host species or from sexually compatible species. The new genes are introduced using recombinant DNA methods and gene transfer. Some scientists hope that the approval process of cisgenic plants might be simpler than that of proper transgenics, but it remains to be seen.

Regulation

The USA is the largest commercial grower of genetically modified crops in the world. For a genetically modified organism to be approved for release it is assessed by the Animal and Plant Health Inspection Service (APHIS) agency within the United States Department of Agriculture (USDA), the Food and Drug Administration (FDA) and the Environmental Protection Agency (EPA).

The USDA evaluated the plants potential to become weeds, the FDA reviewed plants that could enter or alter the food supply and the EPA regulated the genetically modified plants with pesticide properties. Most developed genetically modified plants are reviewed by at least two of the agencies, with many subject to all three. Final approval can still be denied by individual counties within each state. In 2004, Mendocino County, California became the first and only county to impose a ban on the "Propagation, Cultivation, Raising, and Growing of Genetically Modified Organisms", the measure passing with a 57% majority.

The European Union (EU) has possibly the most stringent GMO regulations in the world. All GMOs, along with irradiated food, are considered "new food" and subject to extensive, case-by-case, science based food evaluation by the European Food Safety Authority (EFSA). The EFSA reports to the European Commission who then draft a proposal which if accepted will be adopted by the EC or passed on to the Council of Agricultural Ministers. There is also a safeguard clause that Member States can invoke to restrict or prohibit the use and/or sale of a GMO within their territory if they have a justifiable reasons to consider that the approved GMO constitutes a risk to human health or the environment.

In February 2008 the French government used the safeguard clause to ban the cultivation of MON810 after Senator Jean-François Le Grand, chairman of a committee set up to evaluate biotechnology, said there were "serious doubts" about the safety of the product. In April 2009 German Federal Minister

Ilse Aigner announced an immediate halt to cultivation and marketing of MON810 maize under the safeguard clause.

Currently (2010) the only GMO food crop with approval for cultivation in Europe is the GM maize MON810, which gained approval in 1998. On 2 March 2010 a second GMO, a potato called Amflora, was approved for cultivation for industrial applications in the EU by the European Commission and was grown in Germany, Sweden and the Czech Republic that year. Co-existence of GM and non-GM crops is regulated by the use of buffer zones and isolation distances between the GM and non-GM crops.

The regulations concerning the import and sale of GMOs for human and animal consumption grown outside the EU involve providing freedom of choice to the farmers and consumers. Twice GMOs unapproved by the EC have arrived in the EU and been forced to return to their port of origin. The first was in 2006 when a shipment of rice from America containing an experimental GMO variety (LLRice601) not meant for commercialisation arrived at Rotterdam. The second in 2009 when trace amounts of a GMO maize approved in the US were found in a "non-GM" soy flour cargo.

Genetic engineering in Australia is overseen by the Office of the Gene Technology Regulator (OGTR), a Commonwealth Government Authority within the Department of Health and Ageing. The OGTR reports directly to Parliament through a Ministerial Council on Gene Technology and has legislative powers. The OGTR decides on license applications for the release of all genetically modified organisms and Food Standards Australia New Zealand regulates any GM food. Individual state governments are able to assess the impact of release on markets and trade and apply further legislation to control approved genetically modified productsma for commercial cultivation in 2008. GM canola is grown in Western Australia, while South Australia and Tasmania have extended their moratoriums on all genetically modified crops.

Health Canada, under the Food and Drugs Act, and the Canadian Food Inspection Agency are responsible for evaluating the safety and nutritional value of genetically modified foods. The committee that reviewed the regulations in 2003 was accused by environmental and citizen groups of not representing the full spectrum of public interests and for being too closely aligned to industry groups. In central and South America Mexico, Honduras, Costa Rico, Colombia, Bolivia, Paraguay, Chile, Argentina, Uruguay and Brazil all grow GM crops. In Argentina the National Agricultural Biotechnology Advisory Committee (environmental impact), the National Service of Health and Agrifood Quality (food safety) and the National Agribusiness Direction (effect on trade) assess GM products for release, with the final decision made by the Secretariat of Agriculture, Livestock, Fishery and Food. In Brazil the National Biosafety Technical Commission is responsible for assessing environmental and food safety and prepares guidelines for transport, importation and field experiments involving GM products.

The Council of Ministers evaluates the commercial and economical issues with release. Mexico's senate passed a law allowing planting and selling of genetically modified cotton and soybean in Mexicoin 2005 and in 2009 the government enacted statutory provisions for the regulation of genetically modified maize Mexico is the center of diversity for maize and concerns have been raised about the impact genetically modified maize could have on local strains.

GM crops in China go through three phases of field trials (pilot field testing, environmental release testing, and preproduction testing) before they are submitted to the Office of Agricultural Genetic Engineering Biosafety Administration (OAGEBA) for assessment. Producers must apply to OAGEBA at each stage of the field tests. The Chinese Ministry of Science and Technology developed the first biosafety regulations for GM products in 1993 and they were updated in 2001. Most of the National Biosafety Committee are involved in biotechnology leading to criticisms that they do not represent a wide enough range of public concerns. India regulators cleared the Bt brinjal, a genetically modified eggplant, for commercialisation in October 2009. Following opposition from some scientists, farmers and environmental groups a moratorium was imposed on its release in February 2010. The only other Asian country to currently grow GM crops is the Phillipenes. In 2010, the Common Market for Eastern and Southern Africa (COMESA) proposed a policy where new GM crops would be scientifically assessed by COMESA. If it was demeed safe for the environmental and human health permission would be granted for the crop to be grown in all 19 member countries, although the final decision would be left to each individual country. In 2010 South Africa was the major grower of genetically modified crops in Africa, with smaller areas planted in Burkina Faso and Egypt. Burkina Faso has established a National Biosafety Agency that regulates GM products with advice from various governmental and non-governmental advisory committees.

Kenya passed laws in 2011 which allowed the production and importation of GM crops. The Zambian government rejected a consignment of GMO maize supplied by donors during a famine in 2002 on the basis of the Cartagena Protocol. One of the key issues concerning regulators is whether GM products should be labeled. Labeling can be mandatory up to a threshold GM content level (which varies between countries) or voluntary. A study investigating voluntary labeling in South Africa found that 31% of products labeled as GMO-free had a GM content above 1.0%. In Canada and the USA labeling of GM food is voluntary, while in Europe it all food (including processed food) or feed which contains greater than 0.9% of approved GMOs must be labelled.

Controversy

The examples and perspective in this article may not represent a worldwide view of the subject. Please improve this article and discuss the issue on the talk page.

Biological Process

The use of genetically modified organisms has sparked significant controversy in many areas. Some groups or individuals see the generation and use of GMO as intolerable meddling with biological states or processes that have naturally evolved over long periods of time, while others are concerned about the limitations of modern science to fully comprehend all of the potential negative ramifications of genetic manipulation. Other people see this as a continuation in the role humanity has occupied for thousands of years, modifying the genetics of crops by selecting specimen of crops with the most desirable characteristics as parent for the next generation of crops.

Foodchain

The safety of GMOs in the foodchain has been questioned by some environmental groups, with concerns such as the possibilities that GMOs could introduce new allergens into foods, or contribute to the spread of antibiotic resistance. According to a study published in 1999, there was no current evidence to suggest that the processes used to genetically modify food were inherently harmful. However, a number of more recent studies have raised concern, and environmental groups still discourage consumption in many countries, claiming that GM foods are unnatural and therefore unsafe. Such concerns have led to the adoption of laws and regulations that require safety testing of any new organism produced for human consumption.

GMOs' proponents note that because of the safety testing requirements imposed on GM foods, the risk of introducing a plant variety with a new allergen or toxin using genetic modification is much smaller than using traditional breeding processes. Transgenesis has less impact on the expression of genomes or on protein and metabolite levels than conventional breeding or plant (non-directed) mutagenesis. An example of an allergenic plant created using traditional breeding is the kiwi. One article calculated that the marketing of GM salmon could reduce the cost of salmon by half, thus increasing salmon consumption and preventing 1,400 deaths from heart attack a year in the United States.

Trade in Europe and Africa

In response to negative public opinion, Monsanto announced its decision to remove their seed cereal business from Europe, and environmentalists crashed a World Trade Organization conference in Cancun that promoted GM foods and was sponsored by Committee for a Constructive Tomorrow (CFACT). Some African nations have refused emergency food aid from developed countries, fearing that the food is unsafe. During a conference in the Ethiopian capital of Addis Ababa, Kingsley Amoako, Executive Secretary of the United Nations Economic Commission for Africa (UNECA), encouraged African nations to accept genetically modified food and expressed dissatisfaction in the public's negative opinion of biotechnology.

Agricultural Surpluses

Patrick Mulvany, Chairman of the UK Food Group, accused some governments, especially the Bush administration, of using GM food aid as a way to dispose of unwanted agricultural surpluses. The UN blamed food companies and accused them of violating human rights, calling on governments to regulate these profit-driven firms. It is widely believed that the acceptance of biotechnology and genetically modified foods will also benefit rich research companies and could possibly benefit them more than consumers in underdeveloped nations.

Labeling

While some groups advocate the complete prohibition of GMOs, others call for mandatory labeling of genetically modified food or other products. Other controversies include the definition of patent and property pertaining to products of genetic engineering. According to the documentary Food, Inc. efforts to introduce labeling of GMOs has repeatedly met resistance from lobbyists and politicians affiliated with companies like Monsanto.

Testing

Bruce Stutz's article, "Wanted: GM Seeds for Study," highlights a story of two dozen scientists who spoke out against the research restrictions put forth by companies producing genetically modified (GM) seeds such as DuPont, Monsanto, and Syngenta. In February 2009, after scientists warned the U.S. Environmental protection Agency (EPA) "that industry influence had made independent analyses of transgenic crops impossible," the American Seed Trade Association (ASTA) agreed that they "would allow researchers greater freedom to study the effects of GM food crops." This agreement left many scientists optimistic about the future, but there is little optimism as to whether this agreement has the ability to "alter what has been a research environment rife with obstruction and suspicion."

Impoverished Nations

Some groups believe that impoverished nations will not reap the benefits of biotechnology because they do not have easy access to these developments, cannot afford modern agricultural equipment, and certain aspects of the system revolving around intellectual property rights are unfair to "undeveloped countries". For example, The CGIAR (Consultative Group of International Agricultural Research) is an aid and research organization that has been working to achieve sustainable food security and decrease poverty in undeveloped countries since its formation in 1971. In an evaluation of CGIAR, the World Bank praised its efforts but suggested a shift to genetics research and productivity enhancement. This plan has several obstacles such as patents, commercial licenses, and the difficulty that third world countries have in accessing the international collection of genetic resources and other intellectual property rights that would educate them about modern technology. The International Treaty on Plant Genetic Resources for Food and Agriculture has attempted to remedy this problem, but results have been inconsistent. As a result, "orphan crops", such as teff, millets, cowpeas, and indigenous plants, are important in the countries where they are grown, but receive little investment.

Private Investments

The development and implementation of policies designed to encourage private investments in research and marketing biotechnology that will meet the needs of poverty-stricken nations, increased research on other problems faced by poor nations, and joint efforts by the public and private sectors to ensure the efficient use of technology developed by industrialized nations have been suggested. In addition, industrialized nations have not tested GM technology on tropical plants, focusing on those that grow in temperate climates, even though undeveloped nations and the people that need the extra food live primarily in tropical climates. Some European scientists are concerned that political factors and ideology prevent unbiased assessment of GM technology in some EU countries, with a negative effect on the whole community.

Transgenic Organisms

Another important controversy is the possibility of unforeseen local and global effects as a result of transgenic organisms proliferating. The basic ethical issues involved in genetic research are discussed in the article on genetic engineering. Some critics have raised the concern that conventionally-bred crop plants can be cross-pollinated (bred) from the pollen of modified plants. Pollen can be dispersed over large areas by wind, animals, and insects.

In 2007, the U.S. Department of Agriculture fined Scotts Miracle-Gro $500,000 when modified genetic material from creeping bentgrass, a new golf-course grass Scotts had been testing, was found within close relatives of the same genus (Agrostis) as well as in native grasses up to 21 km (13 mi) away from the test sites, released when freshly cut grass was blown by the wind.

GM proponents point out that outcrossing, as this process is known, is not new. The same thing happens with any new open-pollinated crop variety—newly introduced traits can potentially cross out into neighboring crop plants of the same species and, in some cases, to closely related wild relatives. Defenders of GM technology point out that each GM crop is assessed on a case-by-case basis to determine if there is any risk associated with the outcrossing of the GM trait into wild plant populations.

The fact that a GM plant may outcross with a related wild relative is not, in itself, a risk unless such an occurrence has negative consequences. If, for example, an herbicide resistance trait was to cross into a wild relative of a crop plant it can be predicted that this would not have any consequences except in areas where herbicides are sprayed, such as a farm. In such a setting the farmer can manage this risk by rotating herbicides.

The European Union funds research programs such as Co-Extra, that investigate options and technologies on the coexistence of GM and conventional farming. This also includes research on biological containment strategies and other measures to prevent outcrossing and enable the implementation of coexistence.

If patented genes are outcrossed, even accidentally, to other commercial fields and a person deliberately selects the outcrossed plants for subsequent planting then the patent holder has the right to control the use of those crops. This was supported in Canadian law in the case of Monsanto Canada Inc. v. Schmeiser.

"Terminator" and "Traitor"

An often cited controversy is a "Technology Protection" technology dubbed 'Terminator'. This uncommercialized technology would allow the production of first generation crops that would not generate seeds in the second generation because the plants yield sterile seeds. The patent for this so-called "terminator" gene technology is owned by Delta and Pine Land Company and the United States Department of Agriculture. Delta and Pine Land was bought by Monsanto Company in August 2006. Similarly, the hypothetical trait-specific Genetic Use Restriction Technology, also known as 'Traitor' or 'T-GURT', requires application of a chemical to genetically modified crops to reactivate engineered traits. This technology is intended both to limit the spread of genetically engineered plants, and to require farmers to pay yearly to reactivate the genetically engineered traits of their crops. Genetic Use Restriction Technology is under development by companies including Monsanto and AstraZeneca.

In addition to the commercial protection of proprietary technology in self-pollinating crops such as soybean (a generally contentious issue), another purpose of the terminator gene is to prevent the escape of genetically modified traits from cross-pollinating crops into wild-type species by sterilizing any resultant hybrids. Some environmentalist groups, while considering outcrossing of GM plants dangerous, felt the technology would prevent re-use of seed by farmers growing such terminator varieties in the developing world and was ostensibly a means to exercise patent claims. However other environmental groups welcomed the terminator gene as a means of preventing GM crops from mixing with natural crops.

Hybrid seeds were commonly used in the developed countries long before the introduction of GM crops. Some hybrid crop seeds cannot be saved, so purchasing new seed every year is already a standard agricultural practice for a majority of farms. There are technologies evolving which contain the transgene by biological means and still can provide fertile seeds using fertility restorer functions. Such methods are being developed by several EU research programs, among them Transcontainer and Co-Extra.

BioSteel

BioSteel is a trademark name for a high-strength based fiber material made of the recombinant spider silk-like protein extracted from the milk of transgenic goats, made by Nexia Biotechnologies.

The company has created lines of goats that produce recombinant versions of either the MaSpI or MaSpII dragline silk proteins in their milk. When the female goats lactate, the milk, containing the recombinant silk, is harvested and subjected to chromatographic techniques to purify the recombinant silk proteins. The purified silk proteins are then dried, dissolved using solvents (DOPE formation) and transformed into microfibers using wet-spinning fiber production methodologies. The spun fibers so far have tenacities in the range of 2 - 3 grams/denier and elongation range of 25-45%.

The "Biosteel biopolymer" has been transformed into nanofibers and nano-meshes using the electrospinning technique. Biosteel and other biopolymers are being researched to provide lightweight, strong, and versatile materials for a variety of medical and industrial applications. Nexia Biotechnologies plans to use the spider silk from the milk of transgenic goats for bulletproof vests and anti-ballistic missile systems. Unfortunately, no one has been able to produce it in commercial quantities. Nexia is the only company which has successfully produced fibres from recombinant spider silk and is currently in the process of developing commercial quantities of BioSteel using its transgenic goat technology.

Chimera (genetics)

A chimera or chimaera is a single organism (usually an animal) that is composed of two or more different populations of genetically distinct cells that originated from different zygotes involved in sexual reproduction. If the different cells have emerged from the same zygote, the organism is called a mosaic. Chimeras are formed from four parent cells (two fertilized eggs or early embryos fused together). Each population of cells keeps its own character and the resulting organism is a mixture of tissues. Chimeras are typically seen in animals; there are some reports on human chimerism. Plant chimeras are discussed separately. This condition is either inherited, or it is acquired through the infusion of allogeneic hematopoietic cells during transplantation or transfusion. In nonidentical twins, chimerism occurs by means of blood-vessel anastomoses. The likelihood of offspring being a chimera is increased if it is created via in vitro fertilization. Chimeras can often breed, but the fertility and type of offspring depends on which cell line gave rise to the ovaries or testes; varying degrees of intersexuality may result if one set of cells is genetically female and another genetically male.

Tetragametic Chimerism

Tetragametic chimerism is a form of congenital chimerism. This condition occurs through the fertilization of two separate ova by two sperm, followed by the fusion of the two at the blastocyst or zygote stages. This results in the development of an organism with intermingled cell lines. Put another way, the chimera is formed from the merging of two nonidentical twins (although a similar merging presumably occurs with identical twins, but as their DNA is almost identical, the presence would not be immediately detectable in a very early (zygote or blastocyst) phase. As such, they can be male, female, or hermaphroditic.

As the organism develops, it can come to possess organs that have different sets of chromosomes. For example, the chimera may have a liver composed of cells with one set of chromosomes and have a kidney composed of cells with a second set of chromosomes. This has occurred in humans, and at one time was thought to be extremely rare, though more recent evidence suggests that it is not as rare as previously believed.

This is particularly true for the marmoset. Recent research shows most marmosets are chimerae, sharing DNA with their fraternal twins. Most chimerae will go through life without realizing they are chimeras. The difference in phenotypes may be subtle (e.g., having a hitchhiker's thumb and a straight thumb, eyes of slightly different colors, differential hair growth on opposite sides of the body, etc.) or completely undetectable. Chimera may also show, under a certain spectrum of UV light, distinctive marks on the back resembling that of arrow points pointing downwards from the shoulders down to the lower back; this is one expression of pigment unevenness called

Blaschko's lines. Affected persons may be identified by the finding of two populations of red cells or, if the zygotes are of opposite sex, ambiguous genitalia and hermaphroditism alone or in combination; such persons sometimes also have patchy skin, hair, or eye pigmentation (heterochromia).

If the blastocysts are of opposite sex, genitals of both sexes may be formed, either ovary and testis, or combined ovotestes, in one rare form of intersexuality, a condition previously known as true hermaphroditism.

Note that the frequency of this condition does not indicate the true prevalence of chimerism. Most chimerae composed of both male and female cells probably do not have an intersex condition, as might be expected if the two cell populations were evenly blended throughout the body. Often, most or all of the cells of a single cell type will be composed of a single cell line, i.e. The blood may be composed prominently of one cell line, and the internal organs of the other cell line. Genitalia produce the hormones responsible for other sex characteristics.

If the sex organs are homogeneous, the individual will not be expected to exhibit any intersex traits. Natural chimeras are almost never detected unless they exhibit abnormalities such as male/female or hermaphrodite characteristics or uneven skin pigmentation. The most noticeable are some male tortoiseshell cats or animals with ambiguous sex organs.

The existence of chimerism is problematic for DNA testing, a fact with implications for family and criminal law. The Lydia Fairchild case, for example, was brought to court after DNA testing apparently showed that her children could not be hers. Fraud charges were filed against her and her custody of her children was challenged.

The charge against her was dismissed when it became clear that Lydia was a chimera, with the matching DNA being found in her cervical tissue. Another case was that of Karen Keegan, who was also in danger of losing her children, after a DNA test for a kidney transplant seemed to show she wasn't the mother of her children. The tetragametic state has important implications for organ or stem-cell transplantation. Chimeras typically have immunologic tolerance to both cell lines.

Microchimerism

Microchimerism is the presence of a small number of cells that are genetically distinct from those of the host individual. Apparently, this phenomenon is related to certain types of autoimmune diseases, however, the mechanisms responsible for this relationship are unclear.

"Parasitic" Chimerism in Anglerfish

Chimerism occurs naturally in adult Ceratioid anglerfish and is in fact a natural and essential part of their life cycle. Once a male is born, it begins its search for a female. Using strong olfactory glands, the male searches until it locates a female anglerfish.

The male, less than an inch in length, bites into her skin and releases an enzyme that digests the skin of his mouth and her body, fusing the pair down to the blood-vessel level. While this attachment has become necessary for the male's survival, it will eventually consume him, as both anglerfish fuse into a single hermaphroditic individual. Sometimes in this odd ritual, more than one male will attach to a single female as a 'parasite'.

They will both be consumed into the body of the larger female angler. Once fused to a female, the males will reach sexual maturity, developing large testicles as their other organs atrophy. This process allows for sperm to be in constant supply when the female produces an egg, so that the chimeric fish is able to have a greater number of offspring.

Germline Chimerism

Germline chimerism occurs when the germ cells (for example, sperm and egg cells) of an organism are not genetically identical to its own. It has recently been discovered that marmosets can carry the reproductive cells of their (fraternal) twin siblings, because of placental fusion during development. (Marmosets almost always give birth to fraternal twins.)

Human Chimeras

In 1953 a human chimera was reported in the British Medical Journal. A woman was found to have blood containing two different blood types. Apparently this resulted from cells from her twin brother living in her body. . More recently, a study found that such blood group chimerism is not rare. Another report of a human chimera was published in 1998, where a male human had some partially-developed female organs due to chimerism. He was conceived by in-vitro fertilization.

Chimeras in Research

In biological research, chimeras are artificially produced by selectively transplanting embryonic cells from one organism onto the embryo of another, and allowing the resultant blastocyst to develop. Chimeras are not hybrids, which form from the fusion of gametes from two species that form a single zygote with a combined genetic makeup, or Hybridomas which, as with hybrids, result from fusion of two species' cells into a single cell and artificial propagation of this cell in the laboratory. Essentially, in a chimera, each cell is from either of the parent species, whereas in a hybrid and hybridoma, each cell is derived from both parent species. "Chimera" is a broad term and is often applied to many different mechanisms of the mixing of cells from two different species.

As with cloning, the process of creating and implanting a chimera is imprecise, with the majority of embryos spontaneously terminating. Successes, however, have led to major advancements in the field of embryology, as creating chimeras of one species with different physical traits, such as colour, has allowed researchers to trace the differentiation of embryonic cells through the formation of organ systems in the adult individual. A major milestone in chimera experimentation occurred in 1984, when a chimeric geep was produced by combining embryos from a goat and a sheep, and survived to adulthood. The creation of the "geep" revealed several complexities to chimera development. In implanting a goat embryo for gestation in a sheep, the sheep's immune system would reject the developing goat embryo, whereas a "geep" embryo, sharing markers of immunity with both sheep and goats, was able to survive implantation in either of its parent species.

In August 2003, researchers at the Shanghai Second Medical University in China reported that they had successfully fused human skin cells and dead rabbit eggs to create the first human chimeric embryos. The embryos were allowed to develop for several days in a laboratory setting, then destroyed to harvest the resulting stem cells. In 2007, scientists at the University of Nevada School of Medicine created a sheep whose blood contained 15% human cells and 85% sheep cells. The implications of increasingly realizable projects using human-animal hybrids for biopharmaceutical production, and potentially for producing cells or organs, have raised a host of ethical and safety issues.

Mouse Chimeras

Chimeric mice are important tools in biological research, as they allow the investigation of a variety of biological questions in an animal that has two distinct genetic pools within it. These include insights into such problems as the tissue specific requirements of a gene, cell lineage, and cell potential. The general methods for creating chimeric mice can be summarized either by injection or aggregation of embryonic cells from different origins. The first chimeric mouse was made by Beatrice Mintz in the 1960s through the aggregation of eight cell stage embryos. Injection on the other hand was pioneered by Richard Gardner and Ralph Brinster who injected cells into blastocysts to create chimeric mice with germ lines fully derived from injected ES Cells. Mouse embryos both periimplantation and post implantation can contribute to a chimera.

It is post implantation that ES cells from the inner cell mass of a blastocyst can contribute to all cell lineages of a mouse including the germ line. ES cells are also a useful tool in chimeras because genes can be mutated in them through the use of homologous recombination, thus allowing gene targeting. Since this discovery occurred in 1999, ES cells have become a key tool in the generation of specific chimeric mice.

Underlying Biology

The ability to make mouse chimeras comes from an understanding of early mouse development. Between the stages of fertilization of the egg and the implantation of a blastocyst into the uterus, different parts of the mouse embryo retain the ability to give rise to a variety of cell lineages.

Once the embryo has reached the blastocyst stage, it is composed of several parts, mainly the trophectoderm, the inner cell mass, and the primitive endoderm. Each of these parts of the blastocyst gives rise to different parts of the embryo; the inner cell mass gives rise to the embryo proper, while the trophectoderm and primitive endoderm give rise to extra embryonic structures that support growth of the embryo.

Two- to eight-cell-stage embryos are competent for making chimeras, since at these stages of development, the cells in the embryos are not yet committed to give rise to any particular cell lineage, and could give rise to the inner cell mass or the trophectoderm. In the case where two diploid eight-cell-stage embryos are used to make a chimera, chimersim can be later found in the epiblast, primitive, endoderm and trophectoderm of the mouse blastocyst.

It is possible to dissect the embryo at other stages so as to accordingly give rise to one lineage of cells from an embryo selectively and not the other. For example, subsets of blastomeres can be used to give rise to chimera with specified cell lineage from one embryo. The Inner Cell Mass of a diploid blastocyst for example can be used to make a chimera with another blastocyst of eight-cell diploid embryo; the cells taken from the inner cell mass will give rise to the primitive endoderm and to the epiblast in the chimera mouse.

From this knowledge, ES cell contributions to chimeras have been developed. ES cells can be used in combination with eight-cell-and two-cell-stage embryos to make chimeras and exclusively give rise to the embryo proper. Embryos that are to be used in chimeras can further be genetically altered in order to specifically contribute to only one part of chimera. An example is the chimera built off of ES cells and tetraploid embryos, tetraploid embryos which are artificially made by electrofusion of two two-cell diploid embryos. The tetraploid embryo will exclusively give rise to the trophectoderm and primitive endoderm in the chimera

Methods of Production

There are a variety of combinations that can give rise to a successful chimera mouse and — according to the goal of the experiment — an appropriate cell and embryo combination can be picked; they are generally but not limited to diploid embryo and ES cells, diploid embryo and diploid embryo, ES cell and tetraploid embryo, diploid embryo and tetraploid embryo, ES cells and ES cells. The combination of embryonic stem cell and diploid embryo is a common technique used for the making of chimeric mice, since gene targeting can be done in the embryonic stem cell.

These kinds of chimeras can be made through either aggregation of stem cells and the diploid embryo or injection of the stem cells into the diploid embryo.

If embryonic stem cells are to be used for gene targeting to make a chimera, the following procedure is common: a construct for homologous recombination for the gene targeted will be introduced into cultured mouse embryonic stem cells from the donor mouse, by way of electroporation; cells positive for the recombination event will have antibiotic resistance, provided by the insertion cassette used in the gene targeting; and be able to be positively selected for. ES cells with the correct targeted gene are then injected into a diploid host mouse blastocyst.

These injected blastocysts are then implanted into a pseudo pregnant female surrogate mouse which will bring the embryos to term and give birth to a mouse whose germline is derived from the donor mouse's ES cells. This same procedure can be achieved through aggregation of ES cells and diploid embryos, diploid embryos are cultured in aggregation plates in wells were single embryos can fit, to these wells ES cells are added the aggregates are cultured until a single embryo is formed and has progressed to the blastocyst stage, and can then be transferred to the surrogate mouse.

Legislation

The US and Western Europe have strict codes of ethics and regulations in place that expressly forbid certain subsets of experimentation using human cells, though there is a vast difference in the regulatory framework. In May 2008, a robust debate in the House of Commons of the United Kingdom on the ethics of creating chimeras with human stem cells led to the decision that embryos would be allowed to be made in laboratories, given that they would be destroyed within the first 14 days. No such foundation has been set for chimera research regulation in the US.

Reading Scientific Services (detecting GMOs)

Reading Scientific Services Ltd. (RSSL) is a British company providing scientific analysis, consultancy, product development and training to the food, healthcare, pharmaceutical, personal care, cosmetics and consumer goods industries. It has worked in areas such as the detection of contaminants including dioxins, benzene, melamine, unauthorised colours such as Sudan, undeclared genetically modified organisms, and foreign objects, with an emergency response service for urgent investigations. It also tests pharmaceutical packaging.

The company was formed in 1987 out of Cadbury Schweppes Research and Development. It is a wholly owned subsidiary of Cadbury. It has been inspected by regulatory authorities including the U.S. Food and Drug Administration (FDA), the Medicines and Healthcare products Regulatory Agency (MHRA) and the United Kingdom Accreditation Service (UKAS). RSSL is based at two sites on the Whiteknights Campus of the University of Reading. The headquarters and main laboratories are based at the Reading Science Centre, whilst the training centre and clinical sciences are located at the Science & Technology Centre.

Smart Breeding

SMART breeding (SMART=Selection with Markers and Advanced Reproductive Technologies) or Precision breeding refers to a genetic engineering technique of reproducing a species members together to retain desirable traits and so produce a stronger hybrid. The technique was successfully used by Nachum Kedar, an Israeli scientist, who applied the technique using beefsteak tomatoes to produce a fruit that would ripen on the vine and remain firm in transit. Smart breeding has been advanced as an alternative to transgenics as a way to produce plants that are resistant to various environmental problems.

Sperm-mediated Gene Transfer

Sperm-mediated gene transfer, or SMGT, is a technique used to produce transgenic animals. Genetic material is introduced into sperm, which are used to fertilize eggs. The embryos are carried to term. The offspring may be transgenic, mosaic, or neither.

Synthetic Biology

Synthetic biology is a new area of biological research that combines science and engineering. It encompasses a variety of different approaches, methodologies, and disciplines with a variety of definitions. What they all have in common, however, is that they see synthetic biology as the design and construction of new biological functions and systems not found in nature.

History

The term "synthetic biology" has a history spanning the twentieth century. The first use was in Stéphane Leducs's publication of « Théorie physico-chimique de la vie et générations spontanées » (1910) and « La Biologie Synthétique » (1912). In 1974, the Polish geneticist Waclaw Szybalski used the term "synthetic biology", writing:

Let me now comment on the question "what next". Up to now we are working on the descriptive phase of molecular biology. ... But the real challenge will start when we enter the synthetic biology phase of research in our field. We will then devise new control elements and add these new modules to the existing genomes or build up wholly new genomes. This would be a field with the unlimited expansion potential and hardly any limitations to building "new better control circuits" and ..... finally other "synthetic" organisms, like a "new better mouse". ... I am not concerned that we will run out of exciting and novel ideas, ... in the synthetic biology, in general. When in 1978 the Nobel Prize in Physiology or Medicine was awarded to Arber, Nathans and Smith for the discovery of restriction enzymes, Waclaw Szybalski wrote in an editorial comment in the journal Gene: The work on restriction nucleases not only permits us easily to construct recombinant DNA molecules and to analyze individual genes, but also has led us into the new era of synthetic biology where not only existing genes are described and analyzed but also new gene arrangements can be constructed and evaluated.

Perspectives

Biology

Biologists are interested in learning more about how natural living systems work. One simple, direct way to test our current understanding of a natural living system is to build an instance (or version) of the system in accordance with our current understanding of the system. Michael Elowitz's early work on the Repressilator is one good example of such work. Elowitz had a model for how gene expression should work inside living cells.

To test his model, he built a piece of DNA in accordance with his model, placed the DNA inside living cells, and watched what happened. Slight differences between observation and expectation highlight new science that may be well worth doing. Work of this sort often makes good use of mathematics to predict and study the dynamics of the biological system before experimentally constructing it. A wide variety of mathematical descriptions have been used with varying accuracy, including graph theory, Boolean networks, ordinary differential equations, stochastic differential equations, and Master equations (in order of increasing accuracy). Good examples include the work of Adam Arkin, Jim Collins and Alexander van Oudenaarden. See also the PBS Nova special on artificial life.

Chemistry

Biological systems are physical systems that are made up of chemicals. Around the turn of the 20th century, the science of chemistry went through a transition from studying natural chemicals to trying to design and build new chemicals. This transition led to the field of synthetic chemistry. In the same tradition, some aspects of synthetic biology can be viewed as an extension and application of synthetic chemistry to biology, and include work ranging from the creation of useful new biochemicals to studying the origins of life. Eric Kool's group at Stanford, the Foundation for Applied Molecular Evolution, Carlos Bustamante's group at Berkeley, Jack Szostak's group at Harvard, and David McMillen's group at University of Toronto are good examples of this tradition. Much of the improved economics and versatility of synthetic biology is driven by ongoing improvements in gene synthesis.

Engineering

Engineers view biology as a technology - the systems biotechnology or systems biological engineering . Synthetic Biology includes the broad redefinition and expansion of biotechnology, with the ultimate goals of being able to design and build engineered biological systems that process information, manipulate chemicals, fabricate materials and structures, produce energy, provide food, and maintain and enhance human health and our environment. A good example of these technologies include the work of Chris Voigt, who redesigned the Type III secretion system used by Salmonella typhimurium to secrete spider silk proteins, a strong elastic biomaterial, instead of its own natural infectious proteins.

One aspect of Synthetic Biology which distinguishes it from conventional genetic engineering is a heavy emphasis on developing foundational technologies that make the engineering of biology easier and more reliable. Good examples of engineering in synthetic biology include the pioneering work of Tim Gardner and Jim Collins on an engineered genetic toggle switch, a riboregulator, the Registry of Standard Biological Parts, and the International Genetically Engineered Machine competition (iGEM). Studies in synthetic biology can be subdivided into broad classifications according to the approach they take to the problem at hand: photocell design, biomolecular engineering, genome engineering, and biomolecular-design.

The photocell approach includes projects to make self-replicating systems from entirely synthetic components. Biomolecular engineering includes approaches which aim to create a toolkit of functional units that can be introduced to present new orthogonal functions in living cells. Genome engineering includes approaches to construct synthetic chromosomes for whole or minimal organisms. Biomolecular-design approach refers to the general idea of the de novo design and combination of biomolecular components. The task of each of these approaches is similar: To create a more synthetic entry at a higher level of complexity by manipulating a part of the proceeding level.

Re-writing

Re-writers are Synthetic Biologists who are interested in testing the idea that since natural biological systems are so complicated, we would be better off re-building the natural systems that we care about, from the ground up, in order to provide engineered surrogates that are easier to understand

and interact with. Re-writers draw inspiration from refactoring, a process sometimes used to improve computer software. Drew Endy and his group have done some preliminary work on re-writing (e.g., Refactoring Bacteriophage T7). Oligonucleotides harvested from a photolithographic or inkjet manufactured DNA chip combined with DNA mismatch error-correction allows inexpensive large-scale changes of codons in genetic systems to improve gene expression or incorporate novel amino-acids (see George Church's and Anthony Forster's synthetic cell projects. As in the T7 example above, this favors a synthesis-from-scratch approach.

Key Enabling Technologies

There are several key enabling technologies that are critical to the growth of synthetic biology. The key concepts include standardization of biological parts and hierarchical abstraction to permit using those parts in increasingly complex synthetic systems. Achieving this is greatly aided by basic technologies of reading and writing of DNA (sequencing and fabrication), which are improving in price/performance exponentially (Kurzweil 2001). Measurements under a variety of conditions are needed for accurate modeling and computer-aided-design (CAD).

DNA Sequencing

DNA sequencing is determining the order of the nucleotide bases in a molecule of DNA. Synthetic biologists make use of DNA sequencing in their work in several ways. First, large-scale genome sequencing efforts continue to provide a wealth of information on naturally occurring organisms. This information provides a rich substrate from which synthetic biologists can construct parts and devices. Second, synthetic biologists use sequencing to verify that they fabricated their engineered system as intended. Third, fast, cheap and reliable sequencing can also facilitate rapid detection and identification of synthetic systems and organisms.

Fabrication

A critical limitation in synthetic biology today is the time and effort expended during fabrication of engineered genetic sequences. To speed up the cycle of design, fabrication, testing and redesign, synthetic biology requires more rapid and reliable de novo DNA synthesis and assembly of fragments of DNA, in a process commonly referred to as gene synthesis.

In 2000, researchers at Washington University, mentioned synthesis of the 9.6 kbp Hepatitis C virus genome from chemically synthesized 60 to 80-mersIn 2002 researchers at SUNY Stony Brook succeeded in synthesizing the 7741 base poliovirus genome from its published sequence, producing the second synthetic genome. This took about two years of painstaking work. In 2003 the 5386 bp genome of the bacteriophage Phi X 174 was assembled in about two weeks. In 2006, the same team, at the J. Craig Venter Institute, has constructed and patented a synthetic genome of a novel minimal bacterium, Mycoplasma laboratorium and is working on getting it functioning in a living cell.

In 2007 it was reported that several companies were offering the synthesis of genetic sequences up to 2000 bp long, for a price of about $1 per base pair and a turnaround time of less than two weeks. By September 2009, the price had dropped to less than $0.50 per base pair with some improvement in turn around time. Not only is the price judged lower than the cost of conventional cDNA cloning, the economics make it practical for researchers to design and purchase multiple variants of the same sequence to identify genes or proteins with optimized performance. In 2010, Venter's group announced they had been able to assemble a complete genome of millions of base pairs, insert it into a cell, and cause that cell to start replicating.

Modeling

Models inform the design of engineered biological systems by allowing synthetic biologists to better predict system behavior prior to fabrication. Synthetic biology will benefit from better models of how biological molecules bind substrates and catalyze reactions, how DNA encodes the information needed to specify the cell and how multi-component integrated systems behave. Recently, multiscale models of gene regulatory networks have been developed that focus on synthetic biology applications. Simulations have been used that model all biomolecular interactions in transcription, translation, regulation, and induction of gene regulatory networks, guiding the design of synthetic systems.

Measurement

Precise and accurate quantitative measurements of biological systems are crucial to improving understanding of biology. Such measurements often help to elucidate how biological systems work and provide the basis for model construction and validation. Differences between predicted and measured system behavior can identify gaps in understanding and explain why synthetic systems don't always behave as intended. Technologies which allow many parallel and time-dependent measurements will be especially useful in synthetic biology. Microscopy and flow cytometry are examples of useful measurement technologies.

Examples

This section may require cleanup to meet Wikipedia's quality standards. (Consider using more specific cleanup instructions.) Please help improve this section if you can. The talk page may contain suggestions. (April 2011)

Molecular cloning is a method used frequently by geneticists to obtain large quantities of a particular strand of DNA. It involves shaping a selected piece of DNA and inserting it into the DNA of a bacterium called a plasmid. Once the alien DNA is inserted the bacteria is allowed to replicate thus replicating the DNA that it contains. After replication is completed the copies of foreign DNA are separated from the plasmid. In this sense the bacteria becomes a cyborg because a foreign element is introduced and interacts with the bacteria. The Sleeping Beauty transposon system is an example of an engineered enzyme for inserting precise DNA sequences into genomes of vertebrate animals.

The SB transposon is a synthetic sequence that was created based on deriving a consensus sequence of extinct Tc1/mariner-type transposons that are found as evolutionary relics in the genomes of most, if not all, vertebrates. This enzyme took about a year to engineer and since its creation has been used for gene transfer, gene discovery, and gene therapy applications Biosensor technology is another example of cyborg bacteria.

One such sensor created in Oak Ridge National Laboratory and named "critter on a chip" used a coating of bioluminescent bacteria on a light sensitive computer chip to detect certain petroleum pollutants. When the bacteria sense the pollutant, it lights up and is then processed or amplified.

In Australia, biosensors have been created to detect viruses, bacteria, hormones, drugs, and DNA sequences. In the future scientists hope to create chips that can sense toxins such as environmental estrogens and warfare agents. Even more recently chemists at the University of Nebraska created a humidity gauge by using gold plated bacteria on a silicon chip. With a decrease in humidity there was an increase in the circuit flow. One unique feature that separates the chip from the bioluminescent ones is that that after it has been assimilated the bacteria no longer needs to be kept alive for the humidity gauge to work. Nanotechnology also has made advances by using cyborgs.. Researchers at the École polytechnique de Montréal in Canada have attached a microscopic bead to swimming bacteria. Using a magnetic resonance imaging machine (MRI) the researchers have been able to use the magnetic properties of the bacteria to direct it to certain locations. The bead has no purpose at the moment but researchers hope store drugs or other viral fighting agents inside so that it may be released at the directed location.

Challenges

Opposition to Synthetic Biology

Opposition by civil society groups to Synthetic Biology has been led by the ETC Group who have called for a global moratorium on developments in the field and for no synthetic organisms to be released from the lab. In 2006 38 civil society organizations authored an open letter opposing voluntary regulation of the field and in 2008 ETC Group released the first critical report on the societal impacts of synthetic biology which they dubbed "Extreme Genetic Engineering".

Safety and Security

In addition to numerous scientific and technical challenges, synthetic biology raises questions for ethics, biosecurity, biosafety, involvement of stakeholders and intellectual property.

To date, key stakeholders (especially in the US) have focused primarily on the biosecurity issues, especially the so-called dual-use challenge. For example, while the study of synthetic biology may lead to more efficient ways to produce medical treatments (e.g. against malaria), it may also lead to synthesis or redesign of harmful pathogens (e.g., smallpox) by malicious actors. Proposals for licensing and monitoring the various phases of gene and genome synthesis began to appear in 2004. A 2007 study compared several policy options for governing the security risks associated with synthetic biology.

Other initiatives, such as OpenWetWare, diybio, biopunk, biohack, and possibly others, have attempted to integrate self-regulation in their proliferation of open source synthetic biology projects. However the distributed and diffuse nature of open-source biotechnology may make it more difficult to track, regulate, or mitigate potential biosafety and biosecurity concerns. An initiative for self-regulation has been proposed by the International Association Synthetic Biology that suggests some specific measures to be implemented by the synthetic biology industry, especially DNA synthesis companies. Some scientists, however, argue for a more radical and forward looking approaches to improve safety and security issues. They suggest to use not only physical containment as safety measures, but also trophic and semantic containment. Trophic containment includes for example the design of new and more robust forms of auxotrophy, while semantic containment means the design and construction of completely novel orthogonal life-forms.

Social and Ethical

Online discussion of "societal issues" took place at the SYNBIOSAFE forum on issues regarding ethics, safety, security, IPR, governance, and public perception (summary paper). On July 9–10, 2009, the National Academies' Committee of Science, Technology & Law convened a symposium on "Opportunities and Challenges in the Emerging Field of Synthetic Biology" (transcripts, audio, and presentations available).

Some efforts have been made to engage social issues "upstream" focus on the integral and mutually formative relations among scientific and other human practices. These approaches attempt to invent ongoing and regular forms of collaboration among synthetic biologists, ethicists, political analysts, funders, human scientists and civil society activists. These collaborations have consisted either of intensive, short term meetings, aimed at producing guidelines or regulations, or standing committees whose purpose is limited to protocol review or rule enforcement. Such work has proven valuable in identifying the ways in which synthetic biology intensifies already-known challenges in rDNA technologies. However, these forms are not suited to identifying new challenges as they emerge, and critics worry about uncritical complicity.

An example of efforts to develop ongoing collaboration is the "Human Practices" component of the Synthetic Biology Engineering Research Center in the US and the SYNBIOSAFE project in Europe, coordinated by IDC, that investigated the biosafety, biosecurity and ethical aspects of synthetic biology. A report from the Woodrow Wilson Center and the Hastings Center, a prestigious bioethics research institute, found that ethical concerns in synthetic biology have received scant attention. In January 2009, the Alfred P. Sloan Foundation funded the Woodrow Wilson Center, the Hastings Center, and the J. Craig Venter Institute to examine the public perception, ethics, and policy implications of synthetic biology. Public perception and communication of synthetic biology is the main focus of COSY: Communicating Synthetic Biology, that showed that in the general public synthetic biology is not seen as too different from 'traditional' genetic engineering. To better communicate synthetic biology and its societal ramifications to a broader public, COSY and SYNBIOSAFE published a 38 min. documentary film in October 2009 .

After a series of meetings in the fall of 2010, the Presidential Commission for the study of Bioethical Issues released a report, on December 16, to the President calling for enhanced Federal oversight in the emerging field of synthetic biology. The panel that facilitated the production of the report, composed of 13 scientists, ethicists, and public policy experts, said that the very newness of the science, which involves the design and construction of laboratory-made biological parts, gives regulators, ethicists and others time to identify problems early on and craft solutions that can harness the technology for the public good.

Chapter 4

Nano Genetic Engineering Strategy

Tests for Sex-linked Inheritance

By observing the distribution of the dominant and recessive characters among the offspring of certain matings, a test for determining whether given characters depend on the X-chromosomes is supplied. The test consists in making what are called reciprocal crosses.

- On the one hand mate a dominant female with a recessive male;
- Also mate a recessive female with a dominant male.

Represent a dominant X-chromosome by a capital X, a recessive X-chromosome by a lower-case letter x. In the group of organisms with which we have been dealing, the female has two X-chromosomes, the male but one. The daughters receive an X from each parent, the sons an X from the mother only. A dominant female will be represented by XX, a recessive male by xo; similarly a recessive female is xx, a dominant male is XO.

The two ma tings and their results will then be represented as follows:

- XX by xo gives X + XO Dom Mother Rec. Father Dom. Daughters Dom. Sons
- xx by XO gives X + xo Rec. Mother Dom. Father Dom. Daughters Rec. Sons

Thus when the mother is dominant, the father recessive, all the children are dominant, like themother. When the father is dominant, the mother recessive, the daughters are dominant like the father, the sns recessive like the mother; this is called 'criss-cross inheritance'. Whenever reciprocal crosses give these results, it is certain that the two diverse characteristics (dominant and recessive) result from dfferences in the X-chromosomes of the two parents. 'Criss-cross inheritance' is particularly useful in showing at once that we are dealing with sex-linked inheritace.

In every case where such results are produced, further tests show that the two haracteristics follow in later generations the two different kinds of X-chromosomes wherever they go. t is mainly by the use of these tests that the many different characteristics dependent on diversities in X-chromosomes have been discovered.

Sex-linked Inheritance in Group II

Sex-linked inheritance as we havejust described it was originally discovered n animals belonging to Group I, in which the females have two X-chromosomes, the males but one (with or without a Y- cromosome).

But when the tests we have just described were applied to birds and to certain other animals, it was discovered that there is a scond group (Group II) in which it is the male that has two X-chromosomes, the female but one. In th common fowl, reciprocal crosses were made between 'barred' fowls (Plymouth Rock) and black fowls (Langshan). Here the 'barred condition was found to be dominant.

The results were:

- Barred father by black mother=Barred sons + barred daughters.
- Black father by barred mother = Barred sons + black daughters.

Here the second mating gives 'criss-cross inheritance', showing that we are dealing with sex-linked characters dependent on differences between the X-chromosomes of the parents. But, in the first mating above, all the offspring are like the father instead of like the mother, which latter was the case in the matings. The father in this case has the dominant characteristic (barred), and all the offspring are dominant like the father instead of like the mother.

When the mother is dominant ('barred') the result in the fowls is criss-cross inheritance; while, in the cases before described, criss-cross inheritance occurs when it is the father that is domiant. In fact, in the birds, the males and females simply exchange roles, as compared with their roles in the organisms of Group I. Since the roles of males and females are interchanged in the two groups, their chromosomal conditions must also be interchanged; that is, since in Group I the female has two X's, and in Group II the male plays the same role as the female of Group I, the male of Group II must have two X's, the female but one. The chromosomal conditions in the two group are shown in the above tabulation, X signifying dominant, while x signifies recessive, and o signifies the lack of X (whether Y is present or not).

Careful examination shows that this is the only way in which the results in Group II can be produced. Birds and certain moths show inheritance of the kind typical for Group II. In both these it has since been found under the microscope tha the males have indeed one more chromosome than the females.

Role of Development in X-Chromosomes

A great many characteristics, of various kinds, have been found to be the result of modifications of particular X-chromosomes, and thus to follow in later generations the distribution of the descendants of those chromosomes. Examination of a number of these is desirable, both for their own importance and for the light they throw on the functions of the X-chromosomes. In man, the following characteristics, among others, dependent on diverse types of X-chromosomes, are known from the fact that they show typical sex-linked inheritance.

- *Haemophilia*: Lack of coagulability in the blood. This results from a serious defect in certain X-chromosomes. The existence of defective X-chromosomes having this result shows that the normal X-chromosomes play a part in supplying something necessay for producing normal blood that coagulates properly.
- *Programmes Blindness*: The fact that defectiveness in X-chromosomes causes programmes blindness shows

that the normal X's play a role in producing the normally functioning eyes.

Night Blindness

- Inability to see in a poor light.
- Near-Sightedness ofcertain types.

Progressive atrophy of the muscles. The fact that defects in X-chromosomes have this effect shows that thenormal X's play a role in the normal functioning of the nerves and muscles. A considerable number of other sex-linked characters are known in man. But man is a very unfavourable organism for the study of inheritace. Yet even the little that is known of sex-linked inheritance in man shows that the X-chromosome plays a role in many diverse bodily functions.

For a more complete idea of its role, some organism that can be bred experimentally must be examined. For this urpose the fruit-fly, Drosophila melanogaster, is the best organism to select, since it has been studied more extensively than any other. In the fruit-fly we find peculiarities of the following types that are dependent on alterations in particular X-chromosomes:

- *Many different eye colours*: The normal eye programmes in this organism is a certain shade of red. Defects in different X-chromosomes result in producing, in the individuals that bear these defective chromosomes, many different shades f red, varying from deep red to a very light red, and thence to 'buff', 'ivory, and 'white'. More than a dozen different types of eye programmes are known to result from modifications of the X-chromosome. It is clear that the normal X-chromosome plays an important role in producing the normal eye programmes. Structural peculiarities of the eye, such as 'bar-eye', 'facet eye', 'furrowed eye', and the like, are known to result from modifications of the normal X-chromosome.
- *Wing modifications*: Many different conditions of the wings are known that depend upon diversities among X- chromosomes borne by different individuals. These affect all sorts of features of the wings: size, form, venation, function. The normal X-chromosome obviously

plays an important role in the full and normal development of the wings.

- *Body colours and markings*: Modifications of the X-chromosomes produce the body colours yellow, sable, tan, chrome, lemon, green, and the like, in place of the normal grey. Other changes in X alter the distribution of pigment on the body, giving the characters dot.
- *Body structure*: A defect in X produces irregularities in the abdomen, known as 'abnormal abdomen'.
- *Legs*: A defect in X-chromosomes causes abnormal development of the legs; some of them are wholly or partly reduplicated, so that the total number of legs is increased.
- *Bristles*: Modifications in X-chromosomes result in various different changes in the bristles that are scatered over the body.

Many physiological conditions and functions likewise depend upon the X-chromosomes, since they are change when these chromosomes are modified. Among these are the following:

- *Positive reaction to light*: rosophila with normal X-chromosomes fly towards a source of light. Those having defective X-chromosomes of the kind that produce a tan-coloured odydo not fly towards a source of light.
- *Weakness and short life*: Most of the, defective X-chromosmes that cause structural or other changes in the body (different eye colours, wing forms, and the like) produce likewise weakness and short life. The individuals bearing them are ess resistant to bad conditions, and live or a shorter time than the individuals that bear the nomal X-chromosomes.
- *Life and death*: Some X-chromosomes have defects that are so severe that the individuals bearing them will not live and develop, unless there is resent also a normal X-chromosome. Such defects are known as lethals. The presence of such lethal defects in the X-chromosome has the result that males bearing them die, since they have but one X; while females usually live, since they have an additional X that is often normal.

If a moher that has one lethal X and one normal one is mated to a father that has a normal X, the result is that half of the sons which receive the mother's defective X fail to develop, while the other half of the sons,receiving the mother's normal X, live and develop. All the daughters live, since they all receive a normal X-chromosome from the father. The consequence is that in such families there are twice as many daughters as there are sons. In families of 100 to 200, as occur in Drosophila, this is very striking. X-chromosomes are known to play in other organisms roles similar to those mentioned above for Drosophila, although in no other organism has the matter been so fully studied.

We may summarize what has been brought out above in the following statemets:

- Many defects that appear in individuals are the result of defects in the X-chromosomes that they bear.
- Some of these defects re dominant; they are manifested in all individuals in which the defective X i present.
- But most such defects are recessive; they are manifested only in individuals in which the defective X is the only kind present.
- When both a normal X and a defective X are present, in most cases the normal X performs the required functions, so that the individual is not defective.
- Since the female has two X-chromosomes, while the male has but one, such defects are more frequently manifested in the males. Usually one of the X's present in the female is without the defect, so that she is not defective. But when the male carries a defective X-chromosome, the defect is manifested.
- It is thus advantgeous, so far as the occurrence of defects is concerned, to have two X-chromosomes rather than one.
- This gives the female a considerable advantage over the male in these respects. There are many different types of defects due to defective X-chromosomes; almost all of them are moe common in males than in females. Some of these defects seriously injure the health, or

even result in death, if the defective X is the only kind that is present. It is well known that, in general, the death rate is higher in males than in females; this is probably due to the fact that males have but one X-chromosome.

- Among the different indiiduals of a species are scattered a great number of different types of X-chromosomes having different effects on development. Many of the diverse X-chromosomes are distinctly efective, causing personal defects in the individuals that bear them. Others are not defective.
- Defects or modifications in X-chromosomes affect in different cases al parts and functions of the organism.

The Role f the X-Chromosomes in Development

What do the facts brought out in the preceding sections show as to the phyiological functions of the X-chromosomes; as to their role in deveopment? For every kind of effect resulting from defects or modificacations in the X-chromosomes, there is a corresponding (opposed or diverse)action of normal or unmodified X-chromosomes. Since defective X-chromosomes of a certain type produce in Drosophila short, ill-developed wings, it follows that the normal X-chromosomes act on the development in such a way as to give long, well-developed wings.For if we substitute the norma X for the defective one, this cause the normal well-developed wings to be produced.

Similar reasoning applies to all the defective conditions that result from defective X's; the normal X's act in such a way as to produce the corresponding normal conditions. It follows therefore that the normal X-chromosomes play a role in the production of eye programmes and structure, body programmes and structure, in the development of the legs, the wings, the bristles, in producing normal health, resistance and vigour, and in various physiological processes.

We know, further, that they play a most important part in determining sex, with all that this includes of structural and physiological influence. Clearly therefore the X-chromosomes play a most important part in development. Their action is not limited to any single part of the body, nor to any single class of functions. They enter into the processes of development in such a way as to influence all parts ofthe body, all functions of the body. They begin to influence development very early, as was seen in considering the development of sex. And they play important roles in such relatively late processes as the production of eye colours. It can hardly be doubted that they enter into the developmental activities from practically he beginning, influencing all the later processes that occur.

Autosomes and T-Chromosomes

The X-chromosomes have been dealt with because of the great advantages they offer for experimental study. It was seen that they affect development in many ways, and that in a given species there are many different types of X-chromosomes in the different individuals, causing them to show different characteristics, structural and physiological. Is the X-chromosoe typical in these relations? Shall we find similar effects in the other chromosomes? In addition to the X-chromosomes, there are the autosomes, which are present as a rule as several or many pairs.

In man there are 23 pairs of autosomes: in Drosophila there are three pairs. There is, further, in many species the single Y-chromosome. The autosomes and the Y-chrmosome are distributed from parent to offspring in characteristic ways, differing from the distribution of X-chromosomes. If they affect characteristics, these differencesshould appear in a different system of inheritance. We shall examine first the autosomes.

The Autosomes: Typical Mendelian Inheritance

We have already seen that the autosomes play a role in development, in the fact that they help to determine the sex of the developing individual. In Drosophila, if the number of sets of autosomes, in relation to the number of X's, is changed, this changes the sex. Do autosomes also affect other characteristics? Do the diferent pairs of autosomes have different functions? Are there diverse types f autosomes, with different effects, in different individuals of a species, as is the case in relation to the X-chromosomes? Are there dominant and recessie effects of autosomes, as there are of X? If dominant and recessive characteristics depend on auosomes, then these characteristics must follow the distribution of the autosomes from parents to offspring, as the characteristics ependent on X follow its distribution. We must then examine the method of distribution of the autosomes whether thre are characteristics that show this method.

Method of Distribution of the Autosomes

The ditribution of the autosomes from arents to offspring is much simpler than that of the X-chromosomes.

The main facts are as follows:

- The autsomes are in pairs in all individuals. There are no individuals with unpaired autosomes, like the male of some organisms with relation to X.
- Thus the two sexes are alike as to their autosomes.
- In forming germ cells, one member of each pair of autosomes goes to each germ cell. The sperms and the ova are alike with respect to the autosomes.

  Thus the fundamental rule with relation to the autosomes is that each gem cell carries one member of each pair of autosomes.
- By union of two germ cells the pairs of autosomes are restored.
- Thus each one of the offspring carries in each pair of its autosomes one autosome from each parent.

Now consider a single pair of autosomes, such as the pair marked AA in. Suppose that in different individuals the members of this pair of autosomes produce different effects, as we have seen to be the cae with the X-chromosomes of different individuals. And suppose that, as in the case of X, one of these effects is dominant, the other recessive. How would such characteristics be inherited? This question must be examined with care. For this purpose we may take as an example the different colours found in different varieties of peas, used in the classic experiments of Mendel. The yellow seed programmes of a certain variety is dominant, and results from the constitution of the autosomes of a certain pair. The green seed programmes of another variety is recessive, owing to a different constitution of the autosomes of that same pair.

Call thetwo autosomes of tht pair in the dominant yellow race by the capital letters *AA*, while the two autosomes that give rise to the recessive green programmes in the other race may be called *aa*. The dominant parent may be designated *D*, the recssive parent *R*, When these two parents produce germ cells, each germ cell receives, of course, one chromosome for the pair. The germ cells from the dominant parent *D* have the autosome *A*, those from the recessiveparent *R* have the autosome *a*. The erm cells *A*, from parent *D*, unite each with a germ cell a from parent *R* giving offspring (zygotes) containing both *A* and *a*. These first generation offspring are commonly designated *Fi* (first filial generation). And since the characteristic carried by the autosome *A* is dominant, all these *Fi* offspring manifest the dominant characteristic *D* (yellow, in case yellow-seded peas are crossed with green-seeded peas). Next observe the result of mating together two of these individuals *Aa*, of the *Fi* generation, giving the *Fa* (second filial generation).

Each parent *Aa* gives, according to the general rule, two kinds of germ cells, one kind carrying the dominant autosome *A*, the other the recessive autosome *a*. Each kind of germ cell from one parent unites with each kind from the other. That is, half the *A* germ cells from one parent unite with *A* germ cells from the other parent, half with a germ cells from the other parent; and the same is true for the a germ cells from the first parent.

This gives in equal proportions Fa offspring (zygotes) as follows:

$AA + Aa + a\ A + aa$

But s the two combinations *Aa* and *aA* contain the same chromosomes, the proportions may be written *AA* + 2*Aa-f-aa* The individuals of the constitution AA and Aa contain the dominant chromosome *A*; therefore they manifest the dominant characteristic, *D*. The individuals *aa*, having only the recessive chromosome *a*, manifest the recessive characteristic, *R*. Thus in this F^2 generation, so far as manifested characters are concerned, the proportion of offspring are 3 Dominant to *i* Recessive

Thus if dominant and recessive characteristics depend on the two members of a pair of autosomes, hen a dominant individual *AA* is mated with a recessive individual *aa* the general result is bound to be that the immediate offspring (*Fi*) all manifest the dominant characteristic, while in the *Fa* generation (produced by mating two individuals from *Fi*) there will be three times as many dominant individuals as there are recessives. Now, this is the result that was discovered by Mendel in 1866; it formed the foundation for the working out of Mendelian heredity. The proportion 30 to *iR* in the *Fa* generation is commonly known as 'the Mendelian ratio'; it is one of the most characteristic proportions in this type of inheritance.

Mendel discovered that there are many different characteristics that yield this 3 to *i* ratio. If yellow and green peas are crossed, in the *Fa* generation there are 3 yellows to *i* green; if round peas and wrinkled peas are crossed, in F^2 there are 3 round to *i* wrinkled; and so for may other characteristics in the pea plants with which Mendel worked.

And since the time of Mendel these same ratios have been found for thousands of different characteristics, in great numbers of different organisms, from the pea plant and *D* roso- phila to man. The fact that inheritance gives these ratios is commonly spoken of as Mendel's law, and the type of inheritance which it exemplifies is spoken of as Mendelian inheritance. The type that gives in the *Fa* generation 3*D* to *iR* is the commonest method of inheritance, much commoner than the sex-linked method, in which the characteristics depend on differences in the X-chromosomes.

This is obviously because autosomes are more numerous than X-chromosomes. We find then that inheritance according to Mendel's law takes place exactly as it would if the characteristics concerned depended on the two members of a pair of autosomes. As will be shown later, it can be proved experimentally that such characteristics do indeed depend upon autosomes. Thus the rules of Mendelism are the rules of distribution of the two members of a pair of autosomes, one having a recessive effect, the other a dominant effect. Mendelian heredity is heredity of characteristics that depend on differences in autosomes.

Mendel did not know of the relation of the characteristics to the chromosomes. He discovered, however, certain other important facts about this type of heredity, facts which have become clearly explicable since it has been learned that this type depends on differences in autosomes. These facts must be examined. For this purpose it will be convenient first to define certain terms which are universally employed in dealing with inheritance.

By the union of two germ cells or gametes, as *A* and *a*, there is produced a zygote. If the two germ cells are alike with respect to the chromosomes or characteristics that they carry, the zygote so produced is spoken of as a homozygote; thus *AA* and *aa* are homozygotes. If the two germ cells that unite to form a zygote are unlike, as *A* and *a*, the zygote so produced is called a heterozygote, as *Aa*. The two corresponding characters, one dominant, the other recessive, that are produced by the two differing members of a pair of chromosomes are spoken of s alternative characters, or alleles, or allelomorphs. Any given individual manifests one or the other of the two alternatives, not both. For the alternative characters we shall employ henceforth the term alleles.

The characteristic manifested by a zygote is its phaenotype while its actual constitution is its genotype. Thus, two zygotes *AA* and *Aa* both manifest the dominant character; the dominant character is the phaenotype, or the individuals are said to be phaenotypically dominant. But they are diverse in genotype; one is *AA*, the other *Aa*. In the case of a homo-zygote *AA* or *aa*, the phaenotype and genotype may be held to coincide. Mendel crossed yellow peas with green peas, and found, as before stated, that in the *Fi* generation all the peas were yellow, so that yellow is the dominant allelomorph. In the F^2 there were three times as many yellows as greens. The actual number of seeds in the F^2 generation was in his experiments 8023, of which 6022 were yellow while 2001 were green, so that the ratio of dominants to recessives was, accurately expressed, 3.01 to *i*.

Mendel found that if he bred further the individuals of the F^2 generation, some of these gave different results from others. In the pea plant it is possible to allow each plant to fertilize itself, so that both kinds of germ cells come from the same parent. When this was done with the individuals of the F^2 generation, the following results were observed: The green plants (recessive) gave only recessive green offspring (they 'bred true'). Of the yellow plants (dominant), one-third gave only yellow offspring (they 'bred true'), while two-thirds of them gave both yellow and green offspring, in the proportion of 3 yellows to *i* green. The reason for this is clear when we consider the chromosomes on which the characters depend. If *A* represents the chromosome (autosome) that gives the dominant yellow programmes, while a represents the chromosome that gives the recessive green, the original parents were *AA* and *aa*, the *Fi* generation were all *Aa*, while the *Fa* generation showed the roportions *AA-f* ² *Aa* + *aa*.

The recessive individuals *aa* are homozygotes; teir two chromosomes are alike, being *aa*. When germ cells are produced by these homozygotic individuals, all these germ cells must contain the chromosome *a*, and when two of these unite in self-fertilization, of course they produce in all cases again individuals *aa*. These are, of course, recessives (green), like their parents; that is, the green parents aa 'breed true'.

Similarly the dominant yellows, *AA* (constituting one- third of all the dominants), are homozygotes. Their germ cells all contain the chromosome *A*, and the offspring produced by union of such germ cells, in self-fertilization, are all *AA*, and so doinant yellow like the parents. One-third of all the dominants thus breed true. But the individuals *Aa* are heterozygotic; that is, they have two kinds of chromosomes, *A* and *a*, in the pair.

Thus they produce two kinds of germ cells, *A* and *a*, in equal numbers. And when these germ cells unite, they necessarily give, as we saw on an earlier page, offspring of three kinds, in the proportions *AA* + 2*Aa*4-*aa* Of these, *AA* and 2*Aa* will be dominants (yellow), while *aa* will be recessive (green). Thus the homozygotes *aa*and *AA* breed true, while the heterozygotes, *Aa*, do not.

As Mendel discovered, if the heterozygotes *Aa* are mated with the recessive homozygotes *aa*, they give two kinds of offspring, *Aa* and *aa*; half (*Aa*) being dominant,the other half (*aa*) being recessive. If the heterozygotes *Aa* are mated with the dominant homozygotes *AA*, the offspring *AA* and *Aa* are all dominant, though half (*AA*) are homozyotes, the other half (*Aa*) heterozygotes. All these results are typical for the descendants of two parents different in a single pair of chromosomes, one pair being dominant, the other recessive. For reference it will be well to make a table of these results, characteristic for Mendelian inheritance.

The Results of Mendelian Inheritance

Parents: *AA* and *aa*; both homozygotic. *Fi*: *Aa*; all alike, dominant heterozygotes. *Fs*: *AA* + 2*Aa* + *aa*. Three-fourths of the individuals dominant, onefourth recessive. Half the individuals homozygotes (*AA* and *aa*), the other half heterozygotes (*aAa*). Each type of homozygote bred by itself produces offspring like itself. The heterozygotes *Aa*, when bred by themselves, produce again offspring. *AA* + 2*Aa* + *aa* *Aa* mated with aa yields again *Aa* + *aa* *Aa* mated with *AA* yields again *Aa* + *AA*. All of these results are found to hold true for thousands of different characteristics.

We may therefore summarize the situation in the following general propositions:

- When two different individuals are crossed, in many cases the differences between them follow the same method of distribution as do the two members of a pair of autosomes ýÿone being dominant, the other recessive.
- This distribution is what is known as typical Mendelian heredity.
- The two sexes do not differ in these respects. Each sex has the same proportion of dominants and recessives.
- The rules of Mendelian inheritance are essentially the rules of distribution of the members of pairs of autosomes.

Characteristics that depend on two or more pairs of autosomes.We have thus far followed the distribution of characteristics that depend on differences in a single pair of autosomes. But most organisms have several or many pairs of autosomes: the fruit-fly has three pairs, man has twenty- three. Do some characteristics depend on differences in one pair of autosomes, others on differences in other pairs? Have the different pairs of autosomes thus different functions in development and inheritance? We find that the different pairs of autosomes have indeed different functions, and that some characteristics depend on one pair, some on another. This is shown in many cases i which two parents differ in two diverse characteristics. One characteristic may then act as if dependent on one pair of autosomes, the other on another pair. As an example of this, we may take a case studied by Mendel in peas. One of the original pea plants had seeds that were (i) yellow, and (2) round, or smooth. These, as it turns out, are both dominant characters. The other plants, crossed with those just mentioned, had recessive characters; the seeds were (i) green, and (2) wrinkled. When these two kinds of pea plants were mated together, it turned out, as we shall see, that yellow and green behave in inheritances as if due to a difference in the two parents in the effects of a certain pair of autosomes; that is, they are alleles. Round and wrinkledbehave as if due to a difference between the two parents in the effects of another pair of autosomes.

We may therefore represent the two pairs of autosomes. Representing the dominant characters by capitals, the recessives by lower-case letters, we may in the diagram represent the chromosomes giving origin to the four characters as follows:

A = yellow a = green

B = round b = wrinkled

The germ cells from these parents contain, of course, one member of each pair, as shown in Fig. One parent produces gametes all of the constitution *AB*, the other gametes all of the constitution *ab*. When a gamete *AB* from one parent unites with a gamete *ab* from the other, all the offspring (*Fi*) thus produced have the constitution *AaBa*, as shown at *Fi* in Fig.. Since the dominant autosomes *A* and *B* are both present, all these individuals are domiant for both pairs of characters; that is, they all have seeds that are yellow and round, as Mendel found. But all are heterozygotes-for both of these characters.

Next we mate together two of these *Fi* individuals to give the F^2 generation. In forming gametes the rule is, of course, that each gamete contains one member of each pair of chromosomes. But the two pairs, *Aa* and *Bb*, are distributed to the gametes quite independently. Thus, some get *A* and *B*, some *A* and *b*, some a and *B*, some *a* and *b*. There are thus from each parent four types of gametes, *AB, Ab, aB*, and *ab*, each type being represented by many gametes. The gametes of each of the four types from one parent unite in approximately equal proportions with gametes of each of the four types from the other parent. That is, gametes of the type *AB* unite in different cases with *AB, Ab, aB* or *ab* from the other parent; and so of each of the other gametes of the first parent. The result is to give a set of 16 combinations of gametes (some of which are alike, however, as will be seen). It is not necessary to carry further in our diagram the outlines of the chromosomes.

Inheritance Depending on Two Pairs of Autosomes

AB AB AB AB

AB(i) *Ab*(6) *aB*(7) *ab*(s)

Ab Ab Ab Ab

AB(6) *Ab*(2) *aB*(5) *ab*(8)

aB aB aB aB

AB(7) *Ab*(5) *aB*(3) *ab*(9)

ab ab ab ab

AB(5) *Ab*(8) *aB*(9) *ab*(4)

But, as will be observed, not all these 16 are diverse; thus *ABAb* is the same in constitution or genotype as *AbAB*. If we number the different combinations, and give the same number to those that are alike, as is done (in parentheses) we find that there are nine different combinations or genotypes. That is, two parents that are both *ABab* produce offspring of nine different constitutions or genotypes.

When the parents differed in but one pair of characters there were but 3 different combinations. When there are two pairs the number is 3 > or 9-.

These nine different combinations will manifest the dominant characters *A* or *B*, in all cases where an *A* or *B* is present whether present once, as *A*, or twice, as *AA*); they will manifest the recessive characters *a* or *b* in all cases in which *a* or *b* is present twice, no corresponding capital letter being present. Collecting together these different types, we find the different characteristics to be manifested in the following proportions of the zygotes:

9*AB* + 3*Ab* + *saB* + 1*ab*

That is, 9 out of 16 are dominant for both pairs of characters, 3 are dominant for *A*, recessive for *b*; 3 dominant for *B*, recessive for *a*, and *i* out of 16 is recessive for both. There are thus 4 different phaenotypes.

In Mendel's experiment with peas, in which the two pairs of characters were:

1. Yellow and green;
2. Round and wrinkled; in *Fi* the proportions were 9 that were yellow and round,
3. That were yellow and wrinkled,
4. That were green and round, and *i* that was green and wrinkled.

These proportions hold generally when there are two pairs of characters connected with different pairs of autosomes. It may be observed that the proportions of the individuals of different constitution or genotype in F^2, descended from a cros between parents that differ in respect to two pairs of autosomal characters (so that the parents are *AABB* and *aabb*), can be found directly from the proportions for a single pair of characters. In *Fa*, for the single pairs the proportions are $AA + 2Aa\text{-}haA$ and $BB + 2Bb + bb$ respectively. If now we multiply together algebraically these two polynomials, we obtain the relative proportions for the different combinations produced when the two pairs are together.

These proportions are the following:

$AABB + 2AaBB + aaBB + 2AABb + 4AaBb +$

(1) (7) (3) (6) (5)

$2aaBb + AAbb + 2Aabb + aabb$

(9) (2) (8) (4),

Certain other relations among the nine different types are important, since they hold for all cases of two pairs of characters that are dependent upon different pairs of autosomes.

These are the following:

- There are four combinations (Nos. 1, 2, 3, 4), constituting one-fourth of all the zygotes, that are homozygotes for both pairs of gens. The four homozygotes are diverse in their combinations and in the characters they manifest. Any one of these when self-fertilized or mated with another like itself breeds true: that is, it produces offspring that have the same constitution as the parents.

- Four out of the 16, or one-fourth of all the zygotes, are heterozygotes for both pairs of characters. They form the class numbered (5), and constitute the diagonal from upper right to lower left. Any of these when self-fertilized or mated with another like itself gives again the nine different types, in the proportions above set forth.
- The one-half of all the zygotes, are homozygotes for one of the two pairs, heterozygotes for the other. Four are homozygotic for *A* and heterozygotic for *B*; the other four are heterzygotic for *A* and homozygotic for *B*. In these 8 there are but 4 diverse combinations or genotypes.

If any one of these is self-fertilized, it yields 3 diverse types with respect to the pair for which it is heterozygotic. Thus, the class (6), having the constitution AABb, yields when self-fertilized three combinations in the following proportions:

AABB + 2*AABb* + *AAbb*

The above relations hold for all cases in which two pairs of characteristics are dependent on differences in two different pairs of autosomes, whatever the method of designating the characters or chromosomes. In a similar way, two parents may differ in three characters that are dependent on three different pairs of autosomes. Such parents could be represented as *AABBCC* and *aabbcc*.

Their progeny (*Fi* generation) would all have the constitution *AaBbCc*. Such *Fi* individuals of course produce a variety of germ cells, in accordance with the principles that each germ cell receives one member of each pair of autosomes, and that the different pairs of autosomes are distributed independently. Each parent *AaBbCc* thus yields the following 8 types of germ cells:

ABC, ABc, AbC, aBC, Abe, aBc, abC, abc.

When each of the 8 kinds from one parent unites with each of the 8 kinds from the other, there are produced, of course,

64 groups. If we classify the 64 individuals by the characters which they manifest, we find that there are 8 different phaenotypes manifesting respectively the following characteristics:

ABC, ABc, AbC, aBC, Abc, aBc, abC, and *abc*

That is, some are dominant for all three characters; some for a particular two and recessive for the other; some are dominant for a particular one and recessive for the other two; and some are recessive for all three. The different phaenotypes are present in different proportions; the proportions when a large number are examined are as follows:

2*ABC* + *gABc* + *gAbC* + *gaBC* + *$Abc* + *saBc* + 3*abC* + *iabc*

The total number of different combinations (genotypes) that occur in such a case is.

These combinations can be obtained in their correct proportions by multiplying together algebraically the expressions: Similar but still more complex series are obtained if parents are interbred that differ in four or more characters depending on four or more different pairs of autosomes. If the number of pairs of autosomes is called ↔, the following relations hold:

If parents differ in n autosomes:

- The number of diverse types of gametes produced by the individuals of the *Fi* generation is 2^n.
- The number of different combinations (genotypes) in the zygotes of F^2, formed by mating the 2^n gametes from *Fi*, is s^w.
- The number of different phaenotypes produced in *Fa* is 2^n.

These relations yield very great numbers of diverse combinations, in cases where the number of pairs of autosomes is large. Thus in man, with 23 pairs of autosomes, if all the autosomes differed in the two parents, the number of types of germ cells produced by the *Fi* individuals would be 2, or many millions, the number of possible diverse genotypes in F^2 would be 3, or many billions, and the number of possible different phaenotypes would be 2. (As will appear later, the number of different possible combinations is greatly increased by the exchange of parts by the members of a pair of chromosomes.) Thus typical Mendelian inheritance, dependent on differences between different pairs of autosomes, can be treated to a certain extent mathematically. The individuals in F^2 and later generations occur in proportions represented by simple mathematical ratios. *It will be worth while to formulate the main facts that we have brought out as to autosomal inheritance*:

- Some single pairs of characters, dominant and recessive, follow in inheritance the method of distribution of a single pair of autosomes.
- Some pairs of inherited characters follow the rules of distribution of two or more different pairs of autosomes, in the following particulars:
  - — One member of each pair goes into each germ cell.
  - — The members of the different pairs are distributed to the germ cells independently, so that all possible combinations containing one member of each pair are formed with equal frequency.
  - — Each type of gamete from one parent unites with each type from he other parent, the different matings being equally frequent.
- These methods of distribution result, for the characters manifested in the Fs generation, in certain typical numerical proportions or ratios, known as Mendelian ratios.
- This method of inheritance is known as Mendelism, or Mendelian inheritance. The rules of Mendelism are the result of the rules of distribution of pairs of autosomes.
- Most characteristics of organisms are inherited in this way; many more than are inherited in the sex-linked manner. This is evidently because there is but a single pair of X-chromosomes, while there are several or many pairs of autosomes.
- Thus it is clear that, among the different individuals of a species, there are many different types of autosomes (as there are different types of X-chromosomes), producing many kinds of characteristics.

The T-Chromosome

There remains another chromosome, not included in the autosomes: namely, the Y-chromosome. Has this functions similar to those of X and of the autosomes? Does it too play a role in heredity? As we shall see, the Y-chromosome differs in many ways from the other chromosomes. Let us first summarize the main facts as to the occurrence and distribution of Y-chromosomes.

- Some organisms have no Y-chromosomes. Among those are many insects the dog, the cat, the horse.
- In some organisms it is small and appears to be degenerate. This is the case in many insects; also in man.
- In some organisms the Y-chromosome is large, but of a different form from its mate X. Such is the case in Drosophila
- In some organisms Y is not distinguishable from X inform or size, but experiment shows that it differs from X in function.
- The Y-chromosome descends from father to son only, exclusively in the male line. Normally it never occurs in females.
- Hence any characters that depend on the presence of a particular kind of Y will *(a)* be found in males only, and will *(b)* never be inherited *through* females (since Y does not pass through females).

The T-Chromosome in Drosophila

The functions of the Y- chromosome have been more thoroughly studied in Drosophila than in any other organism. Most of the facts as to this have already been brought out, in connection with our account of the functions of the X-chromosome.

We may summarize as follows the known facts as to the functions of the Y-chromosome in Drosophila:

- The Y-chromosome, although it normally occurs in males, is not required for the production of male individuals. Through non-disjunction, male individuals are produced that contain no Y.

- But the presence of Y is necessary for the fertility of males.
- Y may exist in females, as a result of accidental nondisjunction; it does not affect teir sex.
- During many years of intensive study no characteristics were found that follow the Y-chromosome.
- But recently it has been shown that the presence of Y prevents the appearance of a certain recessive character ('bobbed') that is due to a modified X-chromosome. This modified X results in the production of shortened bristles on the body of the fly. But this peculiarity does not occur if a Y is present, so that Y has the dominant normal allele for 'bobbed'. Indications of certain other ill-defined functions of the Y-chromosome have of late come to view.
- No diverse modifications or defects such as are known in great number for the X-chromosome are known in the Y-chromosome.
- Thus in Drosophila the Y-chromosome appears to have little function; it seems nearly but not quite inert.

Active T-chromosomes in certain Organisms; Relation to Male Secondary Sex Characters. In certain organisms the Y-chromosome is more active and varied in its effects. In such cases it shows in different individuals diverse modifications, comparable to the diverse types of the X-chromosome in Drosophila, and these produce diverse inherited characteristics that appear in males only. This matter is closely bound up with the question of what produces the diversity of secondary sex characters. In many animals there are special characteristics which, like the Y- chromosome, are found only in males. These are the male secondary sex characters: horns, mane, beard, greater size, diversities in form, and the like. Since these appear only in themale, the question arises as to whether they may be due to the presence of the Y-chromosome.

With relation to this the following is the situation:

- In most cases, the presence of male secondary sex characters does ot depend on the presence of the Y-chromosomes.

This is shown by the following:

- There are animals which have no Y-chromosome, yet have male secondary sex characters: for example, the dog, the cat, the horse; also various insects.
- In Drosophila, males that contain no Y (as a result of non-disjunction) have the typical male secodary sex characters.
- In some breeds of sheep the male has horns, while the female has none. t could be suggested that this is due to he fact that the male carries the Y-chromosome, while the female does not. But in other breeds, as Dorsets, both sexes have horns, although only the male has the T-chromosome. And in still other breeds, as Suffolks, neither sex has horns, although here as elsewhere the males have the Y-chromosome.

  If a female of the horned breed (Dorset) is crossed with a male of the hornless breed (Suffolk), in the Fi descendants the males have horns, although they receive their Y-chromosome from the hornless breed. They have inherited the horns through the mother.

  Thus it is clear that in these cases the presence of horns is not due to the presence of the Y-chromosome. Extensive work on the inheritance of horns and horn- lessness in sheep, goats and cattle consistently gives results which show that the presence of horns is not due to the presence of the Y-chromosome, or of any special type of Y-chromosome. In all these cases (a), (b) and (c) the presence of distinctive male secondary sex characters depends in some way on the fact that the male carries but one X-chromosome, and consequently produces the male hormone. Their absence from the female results from the fact that she carries two X-chromosomes, and consequently produces the female hormone.

- But in certain fish, some of the male secondary sex characters apparently do depend on the presence of a

Y of a particular type, for they occur only where that type of Y is present, and they can therefore not be inherited through the mother. That is, the Y-chromosomes of certain individuals or breeds are so modified as to produce certain definite characteristics in the individuals that contain them, while individuals that contain another type of Y do not show these characteristics.

An example of this is found in the small fish Lebistes. In a certain race, the males have a black spot at the base of the dorsal fin, while the females are without the spot. In another race of the same species neither the males nor the females have the black spot. Thus the females of the two races are alike in lacking the black spot, while the males differ. Is this difference the result of differences in the Y-chromosomes of the two kinds of males? This may be tested by crossing members of the two races. In the sptted race, represent the X and Y-chromosome by capital letters; while in the non-spotted race, represent them by the small letters *x* and *y*.

A female of the spotted race, XX, is mated with a male of the non-spotted race, xy, thus:

- Females Males
- Parents XX *xy*
- Non-spotted
- *Fi Xx Xy*
- Non-spotted

The Y-Chromosomes

This produces in Fi females Xx, and males Xy. Neither contains the Y from the spotted race, and neither is spotted. The spot is not inherited through the females of the spotted race.

Next mate the female of the non-spotted race, xx, with a male from the spotted race, XY, as follows:

- Females Males
- Parents xx XY
- Spotted
- Fi xX xY
- Spotted

This yields in Fi females Xx, and males xY. These males carry the Y from the spotted race, and they are spotted. Thus far the spot goes with the Y from the spotted race. This Y may be traced further, for example, by breeding together the non-spotted females xx with the Fi males xY, from the preceding cross, thus:

- Females Males
- Parents x xY
- Spotted
- F2 xx xY
- Spotted

This gives in the Fs females xx, and males xY. These males are again spotted; they carry the Y from the spotted race. Wherever this Y from the spotted race is found, the black spot appears; wherever the Y is from the non-spotted race it does not appear. It appears clear therefore that whether the black spot appears in the male depends upon the type of Y-chromosome that it carries. There are two types of Y-chromosome in the species, one producing the black spot, the other not producing it.

It has been shown by Winge that there are many other programmes markings in this fish that depend for their occurrence on the presence of a particular type of Y-chromosome. They occur only in the males, and only in males that have a certain type of Y. In certain other fish also the Y-chromosomes have been shown to be diverse in different individuals, giving rise to distinctive characteristics in the male.

In a certain plant, Melandrium, also, there are inherited characteristics that depend upon the presence of a particular type of Y-chromosome. There are some indications also that in man the nature of the Y-chromosome affects certain characteristics. In a family described by Schofield, the male parent in the first observed generation had webbed toes.

This was inherited in the male lines only by all the males for four successive generations. It thus followed exactly the distribution of the Y-chromosome from the original male parent. It seems probable therefore that it is due to a defect or modification in the Y-chromosome of that individual. Many other cases of webbed toes or fingers are knon that occur in both sexes, so that they are clearly not due to the Y-chromosome. On the whole, the situation as to the Y-chromosome and its relation to inheritance may be summarized as follows:

- In most organisms Y is a degenerate chromosome, lacking most of the functions manifested by the other chromosomes.
- In ome organisms degeneration has gone so far that the Y-chromosome has completely disappeared.
- In other cases it exists, but is very small.
- But in some species the Y-chromosome has effects on hereditary characters that are similar to those produced by other chromosomes. Different types of inherited characters result in such cases from diversely modified Y-chromosomes.
- Characteristics dependent on the presence of a particular type of Y-chromosome are inherited in the male line exclusively.

Three Main Methods of Distribution of Inherited Characteristics

There exist therefore three diverse systems of distributing hereditary characterstics, depending respectively on the X-chromosomes, the autosomes, and the Y-chromosomes. In each of these the rules of inheritance are given by the special method of distribution of the three types of chromosomes, taken in connection with the facts of dominance and recessiveness. Inheritance which follows the distribution of the X-chromosomes is commonly called sex-linked inheritance. That follwing the distribution of autosomes is known as autosomal or typical Mendelian inheritance. Less important than those just named is Y-chromosome inheritance. All these methods are included under the term Mendelian inheritance, employed in a broad sense. In some organisms still other methods of inheritance exist. In higher animals and plants most inheritance occurs in one of the three methods named in the preceding paragraph.

Characteristics Affected both by Genes and by Environment

Many of the same features that can be altered by changing the genes can likewise be altered by appropriate changes in the environment. Characteristics do not fall into two classes, one exclusively hereditary (or dependent on genes), the other exclusively environmental; but any characteristic is affected both by the materials of which the organism is composed, and by the action of the conditions on these materials. Yet, as we shall see, in practice some characteristics are more readily altered by environmental conditions than are others. Certain characteristics owe most of their peculiarities to diversities among genes. Others are readily affected both by genes and by environment. Still others depend mainly on environmental conditions.

Particularly illuminating are the cases in which characteristics depend in marked degree both on genes and on environmental conditions, so that they are altered by changes in either. A number of well-known cases of this kind will be presented, selecting those that arefrom a technical point of view fully known. In that animal whose genetics is best known, the fruit-fly Drosophila, many of the characteristics that have been studied are defects or abnormalities which are sharply dependent on the presence of particular genes. Such a one is abnormal abdomen.

This appears in the fact that the abdomen is ill formed, the segments not being regular nor sharply marked off. This abnormality is found to be due to a defect in a certain gene of the X-chromosome, so that it shows sex-linked inheritance, the abnormality being dominant. Unless that particular defective gene is present the abdomen is normal. But it is dependent too on the environment.

Individuals with the abnormal gene, if developed in a moist atmosphere, have the abnormality. But if those same individuals are developed in a dry atmosphere, they have normal abdomens. To produce the abnormal abdomen, we must have first the particular kind of defective gene that yields it. But second, even if that is present, we must have, too, the moist atmosphere for development; without it the abnormality is not produced, whatever th gene.

Another character in the fruit-fly that acts in a similar way is what is called reduplicated legs. If a certain defective gene is present in the X-chromosome, the animals show a tendency to produce double legs, or even triple or multiple legs, in place of single ones. Butthis does not happen unless the young develop at a low temperature. To produce reduplicated legs, then, there must be present a certain type of gene, and also a certain extenal condition. Unless both are present, the peculiarity is not produced.

It is clear from these cases that even though an individual inherits the type of gene necessary to produce a certain defect, it may not be inevitable that he should have that defect. A special environment may correct what inheritance leaves imperfect These considerations apply to many thngs in man as well as in other organisms. Another example of the fact that what is produced in development depends both on th genes and on the environment is seen in the production of giants in the fruit-fly. A modified gene, located in the X-chromosome, nea its left end, causes the animals (if they have no other type of X-chromo- somes than this) to become giants; they grow to nearly twice the size of the ordinary fly.

But this increase in size takes place only if the animals having the modified gene are well fed during a certain period of their early lives. If not well fed at this particular time in life, they grow no larger than the usual flies. The giant size requires to produce it a particular type of environment acting on a particular type of gene. If either condition is not fulfilled, giants are not produced.

Again, in plants, the difference between green plants, containing chlorophyll, and white ones, is in some cases due to a difference in genes. Plants with a certain kind of genes will remain white even though grown in the light. In other cases, the difference between white and green plants is not the result of a difference in genes, but of a difference in environment. Plants that live in the dark remain white, even though they contain genes that can produce chlorophyll, while similar plants grown in the light are green.

That is, to produce chlorophyll, a certain type of genes is necessary, and also a certain type of environment. Again, there are red and white varieties of primroses; the difference in programmes is inherited; it is due to gene diversities. But a certain variety produces red flowers when grown in a cool place, white flowers when grown in a warm, moist region, as in a greenhouse. The same diversity in programmes is now due to an environmental difference. Complex interrelations of genetic and environmental action have been described by Emerson with relation to diverse colours in maize plants. In different types, the leaves, the husks, the flowers, and other parts, differ in programmes. Emerson describes and Figs six main types, under each of which are subtypes.

These are the following:

- Purple, with its subtype, weak purple.
- Sun-red, with a subtype, weak sun-red.
- Dilute purple.
- Dilute sun-red.
- Brown.
- Green, with a number of subtypes.

The interrelations of genetic and environmental factors in these colours may be summarized as follows:

- Each programmes and each subtype is constant and hereditary, when the different types are bred under like conditions.
- Each of the colours depends on many different genes, like the eye programmes in Drosophila; they may be altered by changes in any of these genes. The different colours result from different combinations of the genes.
- When the different types are crossed they give typical Mendelian inheritance: in some cases with simple ratios, in others with complex ratios. Thus far the conditions are those for typical hereditary characters; the differences are the result of gene differences, and they follow the distribution of the different types of genes in inheritance. But, also, the colours depend on the conditions to which the plants are subjected while developing. Change of certain conditions changes the colours, as does alteration of the genes. Some of the relations are the following:
- The 'sun-red' type requires exposure to the sun for production of the red programmes. Parts protected from

the sun are green. In the 'green types, on the other hand, subjection to the sun does not cause the production of red programmes. Thus the difference between red and green plants is due in some cases to diversity of genes, as when 'sun-red and 'green types are both grown in the sun. In other cases the difference between red and green plants is due to environmental difference, as when 'sun-red plants grown in the sun are compared with those of the same type grown in the shade. The plants of the 'purple' type do not depend on exposure to the sun for their programmes; they are purplish in programmes whether grown in sun or shade. The degree of programmes produced by exposure to the sun depends on what genes are present. In the typical 'sun-red' variety, exposure to the sun causes a red programmes widely distributed on the plant. In the 'dilute sun-red', exposure to the sun under similar conditions causes only a little red, at the bottom and tips of the leaves, while in the 'green type no red appears. The nature of the soil affects the programmes. 'Dilute sun-red' grown in poor soil and subjected to the sun has almost all parts red; in good soil, with the same relation to the sun, it has very little red. The 'green and 'brown' types are not influenced in their programmes by the nature of the soil.

- Certain particular chemicals in the soil affect the programmes. Presence of much nitrogenous matter tends to suppress the red; if little is present the red programmes is marked.
- Storage of carbohydrates in the leaves of the programmes-producing varieties causes increase of the red programmes. The amount of carbohydrate in the leaves may be increased by experimental procedures, as by removing the growing ears, or by bending a leaf so as to break the conducting vessels; the leaves thereupon become red in the parts with increased carbohydrate. But this effect is not produced in the 'green' types; they do not produce the red programmes even in presence of stored carbohydrates.

It appears clear, therefore, that in maize different genes cause the different varieties to have different metabolic processes, different chemical reactions, some of which cause the production of colours, while others do not. Also, diverse environmental conditions likewise cause changes in these metabolic processes, inducing programmes production in some cases.Whether programmes occurs, and its type and amount, depend on interaction of genetic and environmental conditions.

Another instructive case of the interaction of genes with environmental conditions is furnished by the conditions determining the number of facets in the compound eye of Drosophila. In the normal eye there are present about 800 facets. In certain of the flies a gene at locus 57 in the X-chromosome is so modified as to cause a great reduction in the number of fully developed facets. In this condition, known as Bar-eye, there are only about 80 fully developed facets: these form a 'bar' across the eye. This Bar-eye condition is dominant; the number of facets is reduced even though one of the two genes of the pair is normal. We may call the modified gene that causes the reduction in the number of facets the Bar gene.

The number of fully formed facets, it is found, depends on whether there is present but one or two of the Bar genes, and also on whether there is present in addition a normal gene. In the males (having of course but one X), the presence of the Bar gene reduces the number of normal facets to about 90. In the female, if both the X's contain a Bar gene, the number of facets is about 65. If the female is heterozygotic, so that but one of her X's contains the Bar gene, while the other is normal, the mean number of facets is about 358. Finally, in either male or female, if no Bar gene is present, the number of facets is about 800. Later it was discovered that this Bar gene is sometimes modified in a different way, so as to cause a less marked reduction in the number of facets. This modification is known as infra-Bar. A female containing but one of these infra-Bar genes has about 716 facets; if she contains two she has about 348.

Again, by an extraordinary process, almost or quite unique in known genetics, in some cases two of the Bar genes get united by crossing-over into one X-chromosome. This condition is known as double Bar. When both of the chromosomes carry this double Bar, the number of facets is but about 25. If one carries double Bar, the other single Bar, the number of facets is about 36. Thus the number of facets in the eye depends in many ways on the number and conditions of the genes present. This may be shown in the following diagram, representing the conditions in the two X (or Y) chromosomes of any individual, with the corresponding numbers of eye facets present. In this diagram + = the normal, unmodified X-chromosome Y = the Y-chromosome of the male *B* = the Bar gene *iB* = the infra-Bar gene *BB* = the double Bar condition (two Bar genes in one chromosome) Ghr. *i*: + *iB B iB B iB B BB BB BB BB* Chr.s: + + + *iB* Y *BB* + *iB B BB* Facets 800 716 358 348 98 73 65 45 42 36 25. All these various conditions are inherited as unit-difference, sex-linked characters; they are typical hereditary conditions.

But it is found further that the number of facets in most of these conditions depends also on the temperature to which the developing animals are subjected. The numbers given above are those observed by investigators working with fruit- flies cultivated at ordinary, not precisely controlled, temperatures. If different sets of flies are kept at different temperatures, it is found that high temperatures reduce the number of facets in the individuals that carry the Bar genes, though not in the normal individuals. Krafka found the following mean numbers of facets in the various gene conditions, at different temperatures:

Temperature, Degrees	*Bar Females B-B*	*Bar Males B-Y*	*Double Bar Females BB-BB*	*Double Bar Males BB-Y*
J5	214	270	5i	61
20	122	161	33	37
25	81	121	25‘	28
300131	40	74	‘5	‘4

Thus the number of eye facets depends both on the genes and on the temperature. The number 61 may be produced either by a double Bar Male (*BB*–Y) at 15 degrees or by a Bar Female (*B–B*) at between 25 and 30 degrees. Changing either the genes or the temperature will change the number of facets.

Similar situations occur in vertebrates; one of these may be described briefly. The spotted salamander, Salamandra maculosa, is largely black in programmes, with spots or stripes of yellow. The amount of yellow differs in different cases, so that on the whole some specimens are lighter, some darker. In the lighter individuals the yellow spots tend to run together into rows. The different grades are heritable; certain stocks are regularly darker, others lighter. If darker and lighter individuals are mated, the two conditions show Mendelian inheritance, the darker condition being dominant all the Fi generation are darker, while in Fa there are three dark to one light.

The difference between the two thus depends on a difference in one gene. But it can be shown also that the extent of programmes depends on the environment. If the salamander is kept continuously for a long time (a year or two) on a light background, the yellow spots increase in size: they may run together into stripes or into large yellow areas *B*, so that the animals become much lighter. If, on the other hand, they are kept for long periods on a dark background the yellow areas decrease in size; the animals become on the whole darker. The extent of the yellow programmes thus depends both on the genes and on the environment; it can be altered by appropriately changing either.

In all these cases the following general principles apply:

- What are really inherited (passed on from parent to offspring) are the genes; that is, certain materials that in certain combinations and under certain conditions give rise to certain definite characteristics.
- With the same original genes, different environmental conditions may induce the production of diverse characteristics.
- Also, with the same environmental conditins, different genes may induce the production of different characteristics.
- The same difference in characteristics that is in one case produced by diversity of genes is in other cases produced by diversity of environment.

There can be little doubt that these general statements apply to many characteristics in man. Some combinations of human genes form a individual that is a much better culture medium for the bacteria of tuberculosis than are others. A person who gets such a combination of genes will develop tuberculosis, if he lives in a region in which the germs of that disease are abundant; while another person, with a different set of genes, will not be subject to tuberculosis, even though living under the same conditions. And the person with the genes that make him susceptible to tuberculosis will not have the disease if he prevents infection by the tubercle bacillus.

In relation to such susceptibility there are doubtless a great number of grades, dependent on just what combination of genes is present. Some are extremely susceptible, others less so but still prone to the disease, and so on, up to individuals whose gene combinations are such as to give them natural immunity to tuberculosis.

What then shall be answered when it is asked whether tuberculosis is hereditary or not? One can only say that some gene combinations predispose to it more than others, so that a hereditary factor is involved. But also there is a necessary environmental factor; it is the interrelation of the two sets of factors that determines whether the individual shall be afflicted with the disease. Similar considerations apply to any other disease, acute or chronic. Certain combinations of genes are doubtless more readily attacked by plague, by smallpox, by typhoid, by pneumonia, than others, just as certain genetic constitutions yield more readily to extremes of temperature, to exposure to the elements, to unfit food. There is no affair of life in which the gene combination borne by the individual does not play a part. But the environment too plays a part, often an overwhelmingly important part.

The relative role played by genes on the one hand, and by environment on the other, is not the same for all characteristics. Differences in certain features of the organism are mainly due to differences in the genes of which they are composed. In other characteristics, differences are due mainly to environmental diversities. There are all sorts of intermediate conditions between these extremes. We may pass in review a number of different types of characters from this point of view. Most of the diversities in obvious physical features in most animals are the result of differences in the genes, so that they are mainly matters of inheritance rather than of environment. In Drosophila the forms of the parts of the body, of the legs and wings, the distribution of bristles, the venation of the wings, the colours of body and eyes, are all changeable through the action of diverse genes, but diversities in these features due to environmental action are relatively rare. This is the case also for such characters in the vertebrates, including man. Most of the physical characters of human beings form of their features, complexion, programmes of hair, form of limbs, and the like, as well as sex are settled mainly while the child is carried in the mother's body, when the differences in environment are very slight.

The diversities between individuals in these respects are almost entirely though not quite entirely the result of differences in the genes; they are matters mainly of inheritance. This is demonstrated by what we find in twins. The one-egg twins, having the same genes, are usually closely alike in their physical peculiarities, while the twins derived from different eggs, and so having different genes, are much more diverse with respect to such physical characteristics. At times, of course, one-egg twins become diverse through accidents or diseases which affect one and not the other.

But physical diversities in twins due to genetic differences are much more frequent than those due to environmental differences. In some organisms the gross physical features are readily altered by environmental diversities, so that differences between individuals resulting from the different conditions under which they have lived are common. Such is the case with the habit of growth in higher plants; trees and other plants grown under different conditions are very different in form. In some of the lower animals, such as the hydroids, a similar situation is found.

It is obvious that the details of behaviour are in a high degree under the influence of environmental conditions. The particular acts performed at given moments in any given organism are determined almost entirely by the conditions under which it finds itself. Yet organisms of different genetic constitution behave very differently under effectively the same conditions, and the general pattern of the behaviour of organisms may be largely determined by the genetic constitution.

Major Organs and Tissues Transplanted

Thoracic Organs

- Heart (Deceased-donor only)
- Lung (Deceased-donor and living-related lung transplantation)
- Heart/Lung (Deceased-donor and Domino transplant).

Abdominal Organs

- Kidney (Deceased-donor and Living-Donor)
- Liver (Deceased-donor and Living-Donor)
- Pancreas (Deceased-donor only)
- Intestine (Deceased-donor and Living-Donor)
- Stomach (Deceased-donor only)
- Testis.

Tissues, Cells, Fluids

- Hand (Deceased-donor only),
- Cornea (Deceased-donor only),
- Skin including Face replant (autograft) and Face transplant (extremely rare)
- Islets of Langerhans (Pancreas Islet Cells) (Deceased-donor and Living-Donor)
- Bone marrow/Adult stem cell (Living-Donor and Autograft)
- Blood transfusion/Blood Parts Transfusion (Living-Donor and Autograft)
- Blood vessels (Autograft and Deceased-Donor)
- Heart valve (Deceased-Donor, Living-Donor and Xenograft[Porcine/bovine])
- Bone (Deceased-Donor and Living-Donor).

Types of Donor

Organ donors may be living, or brain dead. Brain dead means the donor must have received an injury (either traumatic or pathological) to the part of the brain that controls heartbeat and breathing. Breathing is maintained via artificial sources, which, in turn, maintains heartbeat. Once brain death has been declared the person can be considered for organ donation. Criteria for brain death vary. Because less than 3% of all deaths in the U.S. are the result of brain death, the overwhelming majority of deaths are ineligible for organ donation, resulting in severe shortages. Tissue may be recovered from donors who are cardiac dead. That is, their breathing and heartbeat has ceased. They are referred to as cadaveric donors. In general, tissues may be recovered from donors up to 24 hours past the cessation of heartbeat. In contrast to organs, most tissues (with the exception of corneas) can be preserved and stored for up to five years, meaning they can be "banked." Also, more than 60 grafts may be obtained from a single tissue donor. Because of these three factors, the ability to recover from a non-heart beating donor, the ability to bank tissue, and the number of grafts available from each donor, tissue transplants are much more common than organ transplants. The American Association of Tissue Banks estimates that more than one million tissue transplants take place in the United States each year.

Living

In "living donors", the donor remains alive and donates a renewable tissue, cell, or fluid (e.g. blood, skin), or donates an organ or part of an organ in which the remaining organ can regenerate or take on the workload of the rest of the organ (primarily single kidney donation, partial donation of liver, small bowel). Regenerative medicine may one day allow for laboratory-grown organs, using patient's own cells via stem cells, or healthy cells extracted from the failing organs.

Deceased

Deceased (formerly cadaveric) are donors who have been declared brain-dead and whose organs are kept viable by ventilators or other mechanical mechanisms until they can be excised for transplantation. Apart from brain-stem dead donors, who have formed the majority of deceased donors for the last twenty years, there is increasing use of Donation after Cardiac Death Donors (formerly non-heart beating donors) to increase the potential pool of donors as demand for transplants continues to grow. These organs have inferior outcomes to organs from a brain-dead donor; however given the scarcity of suitable organs and the number of people who die waiting, any potentially suitable organ must be considered.

Reasons for Donation and Ethical Issues

Living Related Donors

Living related donors donate to family members or friends in whom they have an emotional investment. The risk of surgery is offset by the psychological benefit of not losing someone related to them, or not seeing them suffer the ill effects of waiting on a list.

Paired Exchange

A "paired-exchange" is a technique of matching willing living donors to compatible recipients using serotyping. For example a spouse may be willing to donate a kidney to their partner but cannot since there is not a biological match. The willing spouse's kidney is donated to a matching recipient who also has an incompatible but willing spouse. The second donor must match the first recipient to complete the pair exchange. Typically the surgeries are scheduled simultaneously in case one of the donors decides to back out and the couples are kept anonymous from each other until after the transplant.

Paired exchange programs were popularized in the New England Journal of Medicine article "Ethics of a paired-kidney-exchange program" in 1997 by L.F. Ross. It was also proposed by Felix T. Rapport in 1986 as part of his initial proposals for live-donor transplants "The case for a living emotionally related international kidney donor exchange registry" in *Transplant Proceedings*. A paired exchange is the simplest case of a much larger exchange registry program where willing donors are matched with any number of compatible recipients. Transplant exchange programs have been suggested as early as 1970: "A cooperative kidney typing and exchange program.".

The first pair exchange transplant in the U.S. was in 2001 at Johns Hopkins Hospital. The first complex multihospital kidney exchange involving 12 patients was performed in February 2009 by The Johns Hopkins Hospital, Barnes-Jewish Hospital in St. Louis and Integris Baptist Medical Centre in Oklahoma City.

Another 12-patient multihospital kidney exchange was performed four weeks later by Saint Barnabas Medical Centre in Livingston, New Jersey, Newark Beth Israel Medical Centre and New York-Presbyterian Hospital. Surgical teams led by Johns Hopkins continue to pioneer in this field by having more complex chain of exchange such as eight-way multihospital kidney exchange. In December 2009, a 13 organ 13 recipient matched kidney exchange took place, coordinated through Georgetown University Hospital and Washington Hospital Centre, Washington DC.

Paired-donor exchange, led by work in the New England Program for Kidney Exchange as well as at Johns Hopkins University and the Ohio OPOs may more efficiently allocate organs and lead to more transplants.

Good Samaritan

Good Samaritan or "altruistic" donation is giving a donation to someone not well-known to the donor. Some people choose to do this out of a need to donate. Some donate to the next person on the list; others use some method of choosing a recipient based on criteria important to them. Web sites are being developed that facilitate such donation. It has been featured in recent television journalism that over half of the members of the Jesus Christians, an Australian religious group, have donated kidneys in such a fashion.

Compensated Donation

In compensated donation, donors get money or other compensation in exchange for their organs. This practice is common in some parts of the world, whether legal or not, and is one of the many factors driving medical tourism.

In the United States, The National Organ Transplant Act of 1984 made organ sales illegal. In the United Kingdom, the Human Organ Transplants Act 1989 first made organ sales illegal, and has been superseded by the Human Tissue Act 2004.

In 2007, two major European conferences recommended against the sale of organs.

Recent development of web sites and personal advertisements for organs among listed candidates has raised the stakes when it comes to the selling of organs, and have also sparked significant ethical debates over directed donation, "good-Samaritan" donation, and the current U.S. organ allocation policy. Bioethicist Jacob M. Appel has argued that organ solicitation on billboards and the internet may actually increase the overall supply of organs.

Two books, *Kidney for Sale By Owner* by Mark Cherry (Georgetown University Press, 2005); and *Stakes and Kidneys: Why markets in human body parts are morally imperative* by James Stacey Taylor: (Ashgate Press, 2005); advocate using markets to increase the supply of organs available for transplantation.

In a 2004 journal article Economist Alex Tabarrok argues that allowing organ sales, and elimination of organ donor lists will increase supply, lower costs and diminish social anxiety towards organ markets. Iran has had a legal market for kidneys since 1988, and the market price is of the order of US$1,200 for the recipient. *The Economist* and the Ayn Rand Institute approve and advocate a legal market elsewhere. They argued that if 0.06% of Americans between 19 and 65 were to sell one kidney, the national waiting list would disappear (which, the Economist wrote, happened in Iran). The Economist argued that donating kidneys is no more risky than surrogate motherhood, which can be done legally for pay in most countries.

In Pakistan, 40 percent to 50 percent of the residents of some villages have only one kidney because they have sold the other for a transplant into a wealthy person, probably from another country, said Dr. Farhat Moazam of Pakistan, at a World Health Organisation conference. Pakistani donors are offered $2,500 for a kidney but receive only about half of that because middlemen take so much.

In Chennai, southern India, poor fishermen and their families sold kidneys after their livelihoods were destroyed by the Indian Ocean tsunami on December 26, 2004. About 100 people, mostly women, sold their kidneys for 40,000–60,000 rupees ($900–$1,350). Thilakavathy Agatheesh, 30, who sold a kidney in May 2005 for 40,000 rupees said, "I used to earn some money selling fish but now the post-surgery stomach cramps prevent me from going to work." Most kidney sellers say that selling their kidney was a mistake. In Cyprus in 2010 police closed a fertility clinic under charges of trafficking in human eggs. The Petra Clinic, as it was known locally, imported women from Ukraine and Russia for egg harvesting and sold the genetic material to foreign fertility tourists. This sort of reproductive trafficking violates laws in the European Union. In 2010 the Pulitzer Centre on Crisis Reporting and the magazine Fast Company explored illicit fertility networks in Spain, the United States and Israel.

Allocation of Donated Organs

The overwhelming majority of deceased-donor organs in the United States are allocated by federal contract to the Organ Procurement and Transplantation Network (OPTN), held since it was created by the Organ Transplant Act of 1984 by the United Network for Organ Sharing or UNOS. (UNOS does not handle donor cornea tissue; corneal donor tissue is usually handled by various eye banks.) UNOS allocates organs based on the method considered most fair by the scientific leadership in the field. For kidneys, for instance, that is by waiting time; for livers, it is by MELD (Model of End-Stage Liver Disease), an empirical score based on lab values indicative of the sickness of the patient from liver disease. Experiencing somewhat increased popularity, but still very rare, is directed or targeted donation, in which the family of a deceased donor (often honouring the wishes of the deceased) requests an organ be given to a specific person.

If medically suitable, the allocation system is subverted, and the organ is given to that person. In the United States, there are various lengths of waiting due to the different availabilities of organs in different UNOS regions. In other countries such as the UK, only medical factors and the position on the waiting list can affect who receives the organ. If this is not the desired person, it is noted that this puts them higher on the list.

One of the more publicized cases of this type was the 1994 Chester and Patti Szuber transplant. This was the first time that a parent had received a heart donated by one of their own children. Although the decision to accept the heart from their recently killed child was not an easy decision, the Szuber family agreed that giving Patti's heart to her father would have been something that she would have wanted. Access to organ transplantation is one reason for the growth of medical tourism.

Forced Donation

There have been various accusations that certain authorities are harvesting organs from those the authorities deem undesirable, such as prison populations. The World Medical Association stated that individuals in detention are not in the position to give free consent to donate their organs.

Illegal dissection of corpses is a form of body-snatching and may have taken place to obtain allografts. According to the Chinese Deputy Minister of Health, Huang Jiefu, approximately 95% of all organs used for transplantation are from executed prisoners. The lack of public organ donation program in China is used as a justification for this practice.

However reports in Chinese media raised concerns if executed criminals are the only source for organs used in transplants. In October 2007, bowing to international pressure, the Chinese Medical Association agreed on a moratorium of commercial organ harvesting from condemned prisoners, but did not specify a deadline. China agreed to restrict transplantations from donors to their immediate relatives.

People in other parts of the world are responding to this availability of organs, and a number of individuals (including US and Japanese citizens) have elected to travel to China or India as medical tourists to receive organ transplants which may have been sourced in what might be considered elsewhere to be unethical ways.

Organ Transplantation in Different Countries

Demographics

Despite efforts of international transplantation societies, it is not possible to access an accurate source on the number, rates and outcomes of all forms of transplantation globally; the best that we can achieve is estimations. This is not a sound basis for the future and thus one of the crucial strategies for the Global Alliance in Transplantation is to foster the collection and analysis of global data.

Comparison with DNA

RNA and DNA are both nucleic acids, but differ in three main ways. First, unlike DNA, which is, in general, double-stranded, RNA is a single-stranded molecule in many of its biological roles and has a much shorter chain of nucleotides. Second, while DNA contains *deoxyribose*, RNA contains *ribose* (in deoxyribose there is no hydroxyl group attached to the pentose ring in the 2' position). These hydroxyl groups make RNA less stable than DNA because it is more prone to hydrolysis. Third, the complementary base to adenine is not thymine, as it is in DNA, but rather uracil, which is an unmethylated form of thymine.

Like DNA, most biologically active RNAs, including mRNA, tRNA, rRNA, snRNAs, and other non-coding RNAs, contain self-complementary sequences that allow parts of the RNA to fold and pair with itself to form double helices. Structural analysis of these RNAs has revealed that they are highly structured. Unlike DNA, their structures do not consist of long double helices but rather collections of short helices packed together into structures akin to proteins. In this fashion, RNAs can achieve chemical catalysis, like enzymes. For instance, determination of the structure of the ribosome—an enzyme that catalyzes peptide bond formation—revealed that its active site is composed entirely of RNA.

Structure

Each nucleotide in RNA contains a ribose sugar, with carbons numbered 1' through 5'. A base is attached to the 1' position, in general, adenine (A), cytosine (C), guanine (G), or uracil (U). Adenine and guanine are purines, cytosine, and uracil are pyrimidines. A phosphate group is attached to the 3' position of one ribose and the 5' position of the next. The phosphate groups have a negative charge each at physiological pH, making RNA a charged molecule (polyanion). The bases may form hydrogen bonds between cytosine and guanine, between adenine and uracil and between guanine and uracil. However, other interactions are possible, such as a group of adenine bases binding to each other in a bulge, or the GNRA tetraloop that has a guanine–adenine base-pair. An important structural feature of RNA that distinguishes it from DNA is the presence of a hydroxyl group at the 2' position of the ribose sugar. The presence of this functional group causes the helix to adopt the A-form geometry rather than the B-form most commonly observed in DNA. This results in a very deep and narrow major groove and a shallow and wide minor groove. A second consequence of the presence of the 2'-hydroxyl group is that in conformationally flexible regions of an RNA molecule (that is, not involved in formation of a double helix), it can chemically attack the adjacent phosphodiester bond to cleave the backbone. RNA is transcribed with only four bases (adenine, cytosine, guanine and uracil), but these bases and attached sugars can be modified in numerous ways as the RNAs mature. Pseudouridine (Ø), in which the linkage between uracil and ribose is changed from a C–N bond to a C–C bond, and ribothymidine (T) are found in various places (the most notable ones being in the TØC loop of tRNA). Another notable modified base is hypoxanthine, a deaminated adenine base whose nucleoside is called inosine (I). Inosine plays a key role in the wobble hypothesis of the genetic code.

There are nearly 100 other naturally occurring modified nucleosides, of which pseudouridine and nucleosides with 2'-O-methylribose are the most common. The specific roles of many of these modifications in RNA are not fully understood. However, it is notable that, in ribosomal RNA, many of the post-transcriptional modifications occur in highly functional regions, such as the peptidyl transferase center and the subunit interface, implying that they are important for normal function. The functional form of single stranded RNA molecules, just like proteins, frequently requires a specific tertiary structure. The scaffold for this structure is provided by secondary structural elements that are hydrogen bonds within the molecule. This leads to several recognizable "domains" of secondary structure like hairpin loops, bulges, and internal loops. Since RNA is charged, metal ions such as Mg^{2+} are needed to stabilise many secondary and tertiary structures.

Synthesis

Synthesis of RNA is usually catalyzed by an enzyme—RNA polymerase—using DNA as a template, a process known as transcription. Initiation of transcription begins with the binding of the enzyme to a promoter sequence in the DNA (usually found "upstream" of a gene).

The DNA double helix is unwound by the helicase activity of the enzyme. The enzyme then progresses along the template strand in the 3' to 5' direction, synthesizing a complementary RNA molecule with elongation occurring in the 5' to 3' direction. The DNA sequence also dictates where termination of RNA synthesis will occur. RNAs are often modified by enzymes after transcription. For example, a poly(A) tail and a 5' cap are added to eukaryotic pre-mRNA and introns are removed by the spliceosome. There are also a number of RNA-dependent RNA polymerases that use RNA as their template for synthesis of a new strand of RNA. For instance, a number of RNA viruses (such as poliovirus) use this type of enzyme to replicate their genetic material. Also, RNA-dependent RNA polymerase is part of the RNA interference pathway in many organisms.

Types of RNA

Overview

Messenger RNA (mRNA) is the RNA that carries information from DNA to the ribosome, the sites of protein synthesis (translation) in the cell. The coding sequence of the mRNA determines the amino acid sequence in the protein that is produced. Many RNAs do not code for protein however (about 97% of the transcriptional output is non-protein-coding in eukaryotes).

These so-called non-coding RNAs ("ncRNA") can be encoded by their own genes (RNA genes), but can also derive from mRNA introns. The most prominent examples of non-coding RNAs are transfer RNA (tRNA) and ribosomal RNA (rRNA), both of which are involved in the process of translation.

There are also non-coding RNAs involved in gene regulation, RNA processing and other roles. Certain RNAs are able to catalyse chemical reactions such as cutting and ligating other RNA molecules, and the catalysis of peptide bond formation in the ribosome; these are known as ribozymes.

In Translation

Messenger RNA (mRNA) carries information about a protein sequence to the ribosomes, the protein synthesis factories in the cell. It is coded so that every three nucleotides (a codon) correspond to one amino acid. In eukaryotic cells, once precursor mRNA (pre-mRNA) has been transcribed from DNA, it is processed to mature mRNA. This removes its introns—non-coding sections of the pre-mRNA. The mRNA is then exported from the nucleus to the cytoplasm, where it is bound to ribosomes and translated into its corresponding protein form with the help of tRNA. In prokaryotic cells, which do not have nucleus and cytoplasm compartments, mRNA can bind to ribosomes while it is being transcribed from DNA. After a certain amount of time the message degrades into its component nucleotides with the assistance of ribonucleases.

Transfer RNA (tRNA) is a small RNA chain of about 80 nucleotides that transfers a specific amino acid to a growing polypeptide chain at the ribosomal site of protein synthesis during translation. It has sites for amino acid attachment and an anticodon region for codon recognition that binds to a specific sequence on the messenger RNA chain through hydrogen bonding.

Ribosomal RNA (rRNA) is the catalytic component of the ribosomes. Eukaryotic ribosomes contain four different rRNA molecules: 18S, 5.8S, 28S and 5S rRNA. Three of the rRNA molecules are synthesized in the nucleolus, and one is synthesized elsewhere. In the cytoplasm, ribosomal RNA and protein combine to form a nucleoprotein called a ribosome. The ribosome binds mRNA and carries out protein synthesis. Several ribosomes may be attached to a single mRNA at any time. rRNA is extremely abundant and makes up 80% of the 10 mg/ml RNA found in a typical eukaryotic cytoplasm.

Transfer-messenger RNA (tmRNA) is found in many bacteria and plastids. It tags proteins encoded by mRNAs that lack stop codons for degradation and prevents the ribosome from stalling.

Regulatory RNAs

Several types of RNA can downregulate gene expression by being complementary to a part of an mRNA or a gene's DNA. MicroRNAs (miRNA; 21-22 nt) are found in eukaryotes and act through RNA interference (RNAi), where an effector complex of miRNA and enzymes can break down mRNA to which the miRNA is complementary, block the mRNA from being translated, or accelerate its degradation. While small interfering RNAs (siRNA; 20-25 nt) are often produced by breakdown of viral RNA, there are also endogenous sources of siRNAs. siRNAs act through RNA interference in a fashion similar to miRNAs.

Some miRNAs and siRNAs can cause genes they target to be methylated, thereby decreasing or increasing transcription of those genes. Animals have Piwi-interacting RNAs (piRNA; 29-30 nt) which are active in germline cells and are thought to be a defense against transposons and play a role in gametogenesis.

Many prokaryotes have CRISPR RNAs, a regulatory system similar to RNA interference. Antisense RNAs are widespread; most downregulate a gene, but a few are activators of transcription. One way antisense RNA can act is by binding to an mRNA, forming double-stranded RNA that is enzymatically degraded.

There are many long noncoding RNAs that regulate genes in eukaryotes, one such RNA is Xist, which coats one X chromosome in female mammals and inactivates it. An mRNA may contain regulatory elements itself, such as riboswitches, in the 5' untranslated region or 3' untranslated region; these cis-regulatory elements regulate the activity of that mRNA. The untranslated regions can also contain elements that regulate other genes.

In RNA Processing

Many RNAs are involved in modifying other RNAs. Introns are spliced out of pre-mRNA by spliceosomes, which contain several small nuclear RNAs (snRNA), or the introns can be ribozymes that are spliced by themselves. RNA can also be altered by having its nucleotides modified to other nucleotides than A, C, G and U. In eukaryotes, modifications of RNA nucleotides are generally directed by small nucleolar RNAs (snoRNA; 60-300 nt), found in the nucleolus and cajal bodies. snoRNAs associate with enzymes and guide them to a spot on an RNA by basepairing to that RNA. These enzymes then perform the nucleotide modification. rRNAs and tRNAs are extensively modified, but snRNAs and mRNAs can also be the target of base modification.

RNA Genomes

Like DNA, RNA can carry genetic information. RNA viruses have genomes composed of RNA, and a variety of proteins encoded by that genome. The viral genome is replicated by some of those proteins, while other proteins protect the genome as the virus particle moves to a new host cell. Viroids are another group of pathogens, but they consist only of RNA, do not encode any protein and are replicated by a host plant cell's polymerase.

In Reverse Transcription

Reverse transcribing viruses replicate their genomes by reverse transcribing DNA copies from their RNA; these DNA copies are then transcribed to new RNA. Retrotransposons also spread by copying DNA and RNA from one another, and telomerase contains an RNA that is used as template for building the ends of eukaryotic chromosomes.

Double-stranded RNA

Double-stranded RNA (dsRNA) is RNA with two complementary strands, similar to the DNA found in all cells. dsRNA forms the genetic material of some viruses (double-stranded RNA viruses). Double-stranded RNA such as viral RNA or siRNA can trigger RNA interference in eukaryotes, as well as interferon response in vertebrates.

Key Discoveries in RNA Biology

Research on RNA has led to many important biological discoveries and numerous Nobel Prizes. Nucleic acids were discovered in 1868 by Friedrich Miescher, who called the material 'nuclein' since it was found in the nucleus. It was later discovered that prokaryotic cells, which do not have a nucleus, also contain nucleic acids.

The role of RNA in protein synthesis was suspected already in 1939. Severo Ochoa won the 1959 Nobel Prize in Medicine (shared with Arthur Kornberg) after he discovered an enzyme that can synthesize RNA in the laboratory. Ironically, the enzyme discovered by Ochoa (polynucleotide phosphorylase) was later shown to be responsible for RNA degradation, not RNA synthesis.

The sequence of the 77 nucleotides of a yeast tRNA was found by Robert W. Holley in 1965, winning Holley the 1968 Nobel Prize in Medicine (shared with Har Gobind Khorana and Marshall Nirenberg).

In 1967, Carl Woese hypothesized that RNA might be catalytic and suggested that the earliest forms of life (self-replicating molecules) could have relied on RNA both to carry genetic information and to catalyze biochemical reactions—an RNA world.

During the early 1970s, retroviruses and reverse transcriptase were discovered, showing for the first time that enzymes could copy RNA into DNA (the opposite of the usual route for transmission of genetic information). For this work, David Baltimore, Renato Dulbecco and Howard Temin were awarded a Nobel Prize in 1975.

In 1976, Walter Fiers and his team determined the first complete nucleotide sequence of an RNA virus genome, that of bacteriophage MS2. In 1977, introns and RNA splicing were discovered in both mammalian viruses and in cellular genes, resulting in a 1993 Nobel to Philip Sharp and Richard Roberts.

Catalytic RNA molecules (ribozymes) were discovered in the early 1980s, leading to a 1989 Nobel award to Thomas Cech and Sidney Altman.

In 1990 it was found in petunia that introduced genes can silence similar genes of the plant's own, now known to be a result of RNA interference. At about the same time, 22 nt long RNAs, now called microRNAs, were found to have a role in the development of *C. elegans*. Studies on RNA interference gleaned a Nobel Prize for Andrew Fire and Craig Mello in 2006, and another Nobel was awarded for studies on transcription of RNA to Roger Kornberg in the same year.

The discovery of gene regulatory RNAs has led to attempts to develop drugs made of RNA, such as siRNA, to silence genes.

Transcription and Processing of RNA

Gene Expression: Transcription

The majority of genes are expressed as the proteins they encode. The process occurs in two steps:

- Transcription = DNA → RNA
- Translation = RNA → protein.

Taken together, they make up the "central dogma" of biology: DNA → RNA → protein.

Gene Transcription: DNA → RNA

DNA serves as the template for the synthesis of RNA much as it does for its own replication.

The Steps

- Some 50 different protein transcription factors bind to promoter sites, usually on the 52 side of the gene to be transcribed.
- An enzyme, an RNA polymerase, binds to the complex of transcription factors.
- Working together, they open the DNA double helix.
- The RNA polymerase proceeds to "read" one strand moving in its 32 → 52 direction.
- In eukaryotes, this requires — at least for protein-encoding genes — that the nucleosomes in front of the advancing RNA polymerase (RNAP II) be removed. A complex of proteins is responsible for this. The same complex replaces the nucleosomes after the DNA has been transcribed and RNAP II has moved on.
- As the RNA polymerase travels along the DNA strand, it assembles ribonucleotides (supplied as triphosphates, e.g., ATP) into a strand of RNA.
- Each ribonucleotide is inserted into the growing RNA strand following the rules of base pairing. Thus for each C encountered on the DNA strand, a G is inserted in the RNA; for each G, a C; and for each T, an A. However, each A on the DNA guides the insertion of the pyrimidine uracil (U, from uridine triphosphate, UTP). There is no T in RNA.

Quality control: Occasionally RNA polymerase will select and insert an incorrect, mismatched, ribonucleotide. When this occurs in bacteria (and perhaps in all organisms), the enzyme backs up, removes the incorrect nucleotide (and the one preceding it) and tries again. (Described by Zenkin *et al.*, in the 28 July 2006 issue of Science.)

- Synthesis of the RNA proceeds in the 52 → 32 direction.
- As each nucleoside triphosphate is brought in to add to the 32 end of the growing strand, the two terminal phosphates are removed.
- When transcription is complete, the transcript is released from the polymerase and, shortly thereafter, the polymerase is released from the DNA.

Note that at any place in a DNA molecule, either strand may be serving as the template; that is, some genes "run" one way, some the other (and in a few remarkable cases, the same segment of double helix contains genetic information on both strands!).

In all cases, however, RNA polymerase transcribes the DNA strand in its 32 → 52 direction. A report in the 4 January 2001 issue of Nature shows that RNA polymerase actually tracks around the double helix of DNA. In vitro, at least, when RNA polymerase is immobilized, it spins the DNA molecule around and around as it moves along the molecule. Whether it is the polymerase or the DNA that does the spinning in vivo remains to be determined.

Sedimentation pattern produced by high-speed centrifugation of RNA extracted from the precursors of rabbit red blood cells. The discrete bands represent particular classes of RNA.

The transfer RNAs band at about 4S. The ribosomal RNAs of eukaryotes sediment at 5S, 5.8S, 18S, and 28S. (The larger the sedimentation unit, S, the larger the molecule — but not proportionally.) The RNA forming the band at 9S is the messenger RNA for the synthesis of hemoglobin, the major protein synthesized in these cells. In most types of cells, the messenger RNAs are extremely heterogenous, with small amounts distributed from 6S to 25S.

Several types of RNA are synthesized in the nucleus of eukaryotic cells. Of particular interest are:

- messenger RNA (mRNA). This will later be translated into a polypeptide.

- ribosomal RNA (rRNA). This will be used in the building of ribosomes: machinery for synthesizing proteins by translating mRNA.
- transfer RNA (tRNA). RNA molecules that carry amino acids to the growing polypeptide.
- small nuclear RNA (snRNA). DNA transcription of the genes for mRNA, rRNA, and tRNA produces large precursor molecules ("primary transcripts") that must be processed within the nucleus to produce the functional molecules for export to the cytosol. Some of these processing steps are mediated by snRNAs.
- small nucleolar RNA (snoRNA). These RNAs within the nucleolus have several functions.
- microRNA (miRNA). These are tiny (~22 nucleotides) RNA molecules that appear to regulate the expression of messenger RNA (mRNA) molecules.
- XIST RNA. This inactivates one of the two X chromosomes in female vertebrates.

Messenger RNA (mRNA)

Messenger RNA comes in a wide range of sizes reflecting the size of the polypeptide it encodes. Most cells produce small amounts of thousands of different mRNA molecules, each to be translated into a peptide needed by the cell. Many mRNAs are common to most cells, encoding "housekeeping" proteins needed by all cells (e.g. the enzymes of glycolysis).

Other mRNAs are specific for only certain types of cells. These encode proteins needed for the function of that particular cell (e.g., the mRNA for hemoglobin in the precursors of red blood cells).

Ribosomal RNA (rRNA)

There are 4 kinds. In eukaryotes, these are ;

- 18S rRNA. One of these molecules, along with some 30 different protein molecules, is used to make the small subunit of the ribosome.
- 28S, 5.8S, and 5S rRNA. One each of these molecules, along with some 45 different proteins, are used to make the large subunit of the ribosome.

The S number given each type of rRNA reflects the rate at which the molecules sediment in the ultracentrifuge. The larger the number, the larger the molecule (but not proportionally). The 28S, 18S, and 5.8S molecules are produced by the processing of a single primary transcript from a cluster of identical copies of a single gene. The 5S molecules are produced from a different cluster of identical genes.

Transfer RNA (tRNA)

There are some 32 different kinds of tRNA in a typical eukaryotic cell.

- Each is the product of a separate gene.
- They are small (~4S), containing 73-93 nucleotides.
- Many of the bases in the chain pair with each other forming sections of double helix.
- The unpaired regions form 3 loops.
- Each kind of tRNA carries (at its 32 end) one of the 20 amino acids (thus most amino acids have more than one tRNA responsible for them).
- At one loop, 3 unpaired bases form an anticodon.
- Base pairing between the anticodon and the complementary codon on a mRNA molecule brings the correct amino acid into the growing polypeptide chain.

Small Nuclear RNA (snRNA)

Approximately a dozen different genes for snRNAs, each present in multiple copies, have been identified. The snRNAs have various roles in the processing of the other classes of RNA. For example, several snRNAs are part of the spliceosomes that participate in converting pre-mRNA into mRNA by excising the introns and splicing the exons.

Small Nucleolar RNA (snoRNA)

As the name suggests, these small (60–300 nucleotides) RNAs are found in the nucleolus where they are responsible for several functions:

- Some participate in making ribosomes by helping to cut up the large RNA precursor of the 28S, 18S, and 5.8S molecules.

- Others chemically modify many of the nucleotides in rRNA, tRNA, and snRNA molecules, e.g., by adding methyl groups to ribose.
- Some have been implicated in the alternative splicing of pre-mRNA to different forms of mature mRNA.
- One snoRNA serves as the template for the synthesis of telomeres.

In vertebrates, the snoRNAs are made from introns removed during RNA processing.

Noncoding RNA

Only messenger RNA encodes polypeptides. All the other classes of RNA, including types not mentioned here, are thus called noncoding RNA. Much remains to be learned about the function(s) of some of them.

But, taken together, noncoding RNAs probably account for two-thirds of the transcription going on in the nucleus.

The RNA Polymerases

The RNA polymerases are huge multi-subunit protein complexes. Three kinds are found in eukaryotes.

- RNA polymerase I (Pol I). It transcribes the rRNA genes for the precursor of the 28S, 18S, and 5.8S molecules (and is the busiest of the RNA polymerases).
- RNA polymerase II (Pol II; also known as RNAP II). It transcribes protein-encoding genes into mRNA (and also the snRNA genes).
- RNA polymerase III (Pol III). It transcribes the 5S rRNA genes and all the tRNA genes.

Two remarkable reports in the 8 June 2001 issue of Science show the structure of Pol II and reveal many details of how it uses the antisense strand of DNA to synthesize a strand of mRNA.

RNA Processing: Pre-mRNA '! mRNA

All the primary transcripts produced in the nucleus must undergo processing steps to produce functional RNA molecules for export to the cytosol. We shall confine ourselves to a view of the steps as they occur in the processing of pre-mRNA to mRNA. Most eukaryotic genes are split into segments.

In decoding the open reading frame of a gene for a known protein, one usually encounters periodic stretches of DNA calling for amino acids that do not occur in the actual protein product of that gene. Such stretches of DNA, which get transcribed into RNA but not translated into protein, are called introns. Those stretches of DNA that do code for amino acids in the protein are called exons. Examples:

- The gene for one type of collagen found in chickens is split into 52 separate exons.
- The gene for dystrophin, which is mutated in boys with muscular dystrophy, has 79 exons.
- Even the genes for rRNA and tRNA are split by introns.
- The human genome is estimated to contain some 180,000 exons. With a current estimate of 21,000 genes, the average exon content of our genes is about 9.

In general, introns tend to be much longer than exons. An average eukaryotic exon is only 140 nucleotides long, but one human intron stretches for 480,000 nucleotides! Removal of the introns — and splicing the exons together — are among the essential steps in synthesizing mRNA.

The Steps of RNA Processing

- Synthesis of the cap. This is a modified guanine (G) which is attached to the 52 end of the pre-mRNA as it emerges from RNA polymerase II (RNAP II). The cap
  - o protects the RNA from being degraded by enzymes that degrade RNA from the 52 end;
  - o serves as an assembly point for the proteins needed to recruit the small subunit of the ribosome to begin translation.
- Step-by-step removal of introns present in the pre-mRNA and splicing of the remaining exons. This step takes place as the pre-mRNA continues to emerge from RNAP II.
- Synthesis of the poly(A) tail. This is a stretch of adenine (A) nucleotides. When a special poly(A) attachment site

in the pre-mRNA emerges from RNAP II, the transcript is cut there, and the poly(A) tail is attached to the exposed 32 end. This completes the mRNA molecule, which is now ready for export to the cytosol. (The remainder of the transcript is degraded, and the RNA polymerase leaves the DNA).

The cutting and splicing of mRNA must be done with great precision. If even one nucleotide is left over from an intron or one is removed from an exon, the reading frame from that point on will be shifted, producing new codons specifying a totally different sequence of amino acids from that point to the end of the molecule (which often ends prematurely anyway when the shifted reading frame generates a STOP codon).

The removal of introns and splicing of exons is done by spliceosomes. These are a complexes of 5 snRNA molecules and some 145 different proteins. The introns in most pre-mRNAs begin with a GU and end with an AG. Presumably these short sequences assist in guiding the spliceosome.

Visual Evidence

The upper image is an electron micrograph of a mRNA-DNA hybrid molecule formed by mixing the messenger RNA (mRNA) from a clone of antibody-secreting cells with single-stranded DNA from the same kind of cells. The bar represents the length of 1000 bases. The lower diagram is an interpretation of the micrograph. The solid line represents the DNA; the dotted line the mRNA. The loops (I_A, I_B, etc.) represent the introns that separate the exons encoding the domains of an antibody heavy chain:

- V = variable region
- E_1 = first constant region (C_H1) domain
- E_H = hinge region
- E_2 and E_3 = the nucleotides encoding the two C-terminal domains (C_H2 and C_H3).

The unhybridized portion of the mRNA is its poly(A) tail.

Alternative Splicing

The processing of pre-mRNA for many proteins proceeds along various paths in different cells or under different conditions. For example, early in the differentiation of a B cell (a lymphocyte that synthesizes an antibody) the cell first uses an exon that encodes a transmembrane domain that causes the molecule to be retained at the cell surface. Later, the B cell switches to using a different exon whose domain enables the protein to be secreted from the cell as a circulating antibody molecule.

Alternative splicing provides a mechanism for producing a wide variety of proteins from a small number of genes. While we humans may turn out to have only some 20 thousand genes, we probably make at least 10 times that number of different proteins.

It is now estimated that 92–94% of our genes produce pre-mRNAs that are alternatively-spliced. There is evidence that the pattern of alternative splicing differs consistently in different tissues and so must be regulated. But whether all the products are functional or that many are simply the outcome of an error-prone process remains to be seen.

Alternative splicing not only provides different proteins from a single gene but also different 3' UTRs and 5' UTRs. Although not translated into protein, these untranslated regions contain signals that, for example, dictate where in the cell that protein will accumulate. Two examples:

- The 3' UTR of the *bicoid* gene in Drosophila directs the mRNA to the anterior of the embryo;
- the same region in the *VegT* gene of Xenopus directs its mRNA to the vegetal pole of the embryo.

One of the most dramatic examples of alternative splicing is the *Dscam* gene in Drosophila. This single gene contains some 116 exons of which 17 are retained in the final mRNA. Some exons are always included; others are selected from an array.

Theoretically this system is able to produce 38,016 different proteins. And, in fact, over 18,000 different ones have been found in Drosophila hemolymph. These Dscam proteins are used to establish a unique identity for each neuron. It works like this. Each developing neuron draws upon the pool of thousands of possible different mRNAs to synthesize a dozen or so of them. Which ones are selected appears to be simply a matter of chance, but because of the size of the pool, each neuron will most likely end up with a unique set of a dozen or so Dscam proteins. As each developing neuron in the central nervous system sprouts dendrites and an axon, these express its unique collection of Dscam proteins. If the extensions of a single neuron should meet each other in the tangled web that is the hallmark of nervous tissue, they are repelled. In this way, thousands of different neurons can coexist in intimate contact without the danger of nonfunctional contacts between the various extensions of the same neuron. We also have a *DSCAM* gene which is located on chromosome 21. When present in 3 copies, it is probably responsible for some of the features of Down syndrome (and accounts for its full name: Down Syndrome Cell Adhesion Molecule). Perhaps the incredible diversity of synaptic junctions in the mammalian c.n.s. ($\sim 10^{14}$) is mediated by alternative splicing of a limited number of gene transcripts.

So, whether a particular segment of RNA will be retained as an exon or excised as an intron can vary under different circumstances, such as;

- what type of cell the gene is in;
- what stage of differentiation that cell is passing through;
- what extracellular signals that cell is receiving.

Clearly the switching to an alternate splicing pathway must be closely regulated.

Trans-splicing

Most genes are transcribed and their transcripts processed as described above. RNA polymerase travels down a single strand of a singe gene locus to form pre-mRNA that is processed (including removal of introns) to form the mature mRNA. But there are exceptions. A number of cases have been found where two different precursor transcripts have been spliced together to form the final RNA molecule. The phenomenon is called *trans*-splicing.

Examples: synthesis of a single RNA molecule by splicing together transcripts from ;

- loci located far apart on the same chromosome;
- opposite strands of the same gene locus;

- the two alleles of the gene on their separate (homologous) chromosomes.

The biological importance of these *trans*-spliced transcripts is still unknown for most of them.

Why Split Genes?

Perhaps during evolution, eukaryotic genes have been assembled from smaller, primitive genes - today's exons. Some proteins, like the antibodies mentioned in the previous section, are organized in a set of separate sections or domains each with a special function to perform in the complete molecule.

Each domain is encoded by a separate exon. Having the different functional parts of the antibody molecule encoded by separate exons makes it possible to use these units in different combinations.

Thus a set of exons in the genome may be the genetic equivalent of the various modular pieces in a box of "Lego" for children to assemble in whatever forms they wish. But the boundaries of other exons do not seem to correspond to domain boundaries of the protein. So perhaps the major benefit of split genes is simply the opportunity they provide for making many different proteins from a single gene through alternative splicing.

Summary

Gene expression occurs in two steps:

- transcription of the information encoded in DNA into a molecule of RNA (described here) and
- translation of the information encoded in the nucleotides of mRNA into a defined sequence of amino acids in a protein (discussed in Gene Translation: RNA $\rightarrow$ Protein).

Messenger RNA

Messenger RNA (mRNA) is a molecule of RNA encoding a chemical "blueprint" for a protein product. mRNA is transcribed from a DNA template, and carries coding information to the sites of protein synthesis: the ribosomes. Here, the nucleic acid polymer is translated into a polymer of amino acids: a protein. In mRNA as in DNA, genetic information is encoded in the sequence of nucleotides arranged into codons consisting of three bases each. Each codon encodes for a specific amino acid, except the stop codons that terminate protein synthesis. This process requires two other types of RNA: transfer RNA (tRNA) mediates recognition of the codon and provides the corresponding amino acid, while ribosomal RNA (rRNA) is the central component of the ribosome's protein manufacturing machinery.

Synthesis, Processing, and Function

The brief existence of an mRNA molecule begins with transcription and ultimately ends in degradation. During its life, an mRNA molecule may also be processed, edited, and transported prior to translation. Eukaryotic mRNA molecules often require extensive processing and transport, while prokaryotic molecules do not.

Transcription

During transcription, RNA polymerase makes a copy of a gene from the DNA to mRNA as needed. This process is similar in eukaryotes and prokaryotes. One notable difference, however, is that prokaryotic RNA polymerase associates with mRNA processing enzymes during transcription so that processing can proceed quickly after the start of transcription. The short-lived, unprocessed or partially processed, product is termed *pre-mRNA*; once completely processed, it is termed *mature mRNA*.

Eukaryotic Pre-mRNA Processing

Processing of mRNA differs greatly among eukaryotes, bacteria and archea. Non-eukaryotic mRNA is essentially mature upon transcription and requires no processing, except in rare cases. Eukaryotic pre-mRNA, however, requires extensive processing.

5' Cap Addition

A *5' cap* (also termed an RNA cap, an RNA 7-methylguanosine cap or an RNA m^7G cap) is a modified guanine nucleotide that has been added to the "front" or 5' end of a eukaryotic messenger RNA shortly after the start of transcription. The 5' cap consists of a terminal 7-methylguanosine residue which is linked through a 5'-5'-triphosphate bond to the first transcribed nucleotide. Its presence is critical for recognition by the ribosome and protection from RNases. Cap addition is coupled to transcription, and occurs co-transcriptionally, such that each influences the other. Shortly after the start of transcription, the 5' end of the mRNA being synthesized is bound by a cap-synthesizing complex associated with RNA polymerase. This enzymatic complex catalyzes the chemical reactions that are required for mRNA capping. Synthesis proceeds as a multi-step biochemical reaction.

Splicing

Splicing is the process by which pre-mRNA is modified to remove certain stretches of non-coding sequences called introns; the stretches that remain include protein-coding sequences and are called exons. Sometimes pre-mRNA messages may be spliced in several different ways, allowing a single gene to encode multiple proteins.

This process is called alternative splicing. Splicing is usually performed by an RNA-protein complex called the spliceosome, but some RNA molecules are also capable of catalyzing their own splicing.

Ribozyme

A ribozyme (from ribonucleic acid enzyme, also called RNA enzyme or catalytic RNA) is an RNA molecule possessing a well defined tertiary structure that enables it to catalyze a chemical reaction. Many natural ribozymes catalyze either the hydrolysis of one of their own phosphodiester bonds (self-cleaving ribozymes), or the hydrolysis of bonds in other RNAs, but they have also been found to catalyze the aminotransferase activity of the ribosome.

Investigators studying the origin of life have produced ribozymes in the laboratory that are capable of catalyzing their own synthesis under very specific conditions, such as an RNA polymerase ribozyme. Mutagenesis and selection has been performed resulting in isolation of improved variants of the “Round-18” polymerase ribozyme from 2001. “B6.61” is able to add up to 20 nucleotides to a primer template in 24 hours, until it decomposes by hydrolysis of its phosphodiester bonds.

Some ribozymes may play an important role as therapeutic agents, as enzymes which tailor defined RNA sequences, as biosensors, and for applications in functional genomics and gene discovery.

Activity

Although most ribozymes are quite rare in the cell, their roles are sometimes essential to life. For example, the functional part of the ribosome, the molecular machine that translates RNA into proteins, is fundamentally a ribozyme, composed of RNA tertiary structural motifs that are often coordinated to metal ions such as Mg^{2+} as cofactors. There is no requirement for divalent cations in a five-nucleotide RNA that can catalyze *trans*-phenylalanation of a four-nucleotide substrate which has three base complementary sequence with the catalyst. The catalyst and substrate were devised by truncation of the C3 ribozyme.

RNA can also act as a hereditary molecule, which encouraged Walter Gilbert to propose that in the distant past, the cell used RNA as both the genetic material and the structural and catalytic molecule, rather than dividing these functions between DNA and protein as they are today. This hypothesis became known as the "RNA world hypothesis" of the origin of life. If ribozymes were the first molecular machines used by early life, then today's remaining ribozymes — such as the ribosome machinery — could be considered living fossils of a life based primarily on nucleic acids. A recent test-tube study of prion folding suggests that an RNA may catalyze the pathological protein conformation in the manner of a chaperone enzyme.

Known Ribozymes

Naturally occurring ribozymes include:

- Peptidyl transferase 23S rRNA
- RNase P
- Group I and Group II introns
- GIR1 branching ribozyme
- Leadzyme - Although initially created *in vitro*, natural examples have been found
- Hairpin ribozyme
- Hammerhead ribozyme
- HDV ribozyme
- Mammalian CPEB3 ribozyme

- VS ribozyme
- *glmS* ribozyme
- CoTC ribozyme.

Artificial Ribozymes

Since the discovery of ribozymes that exist in living organisms, there has been interest in the study of new synthetic ribozymes made in the laboratory. For example, artificially-produced self-cleaving RNAs that have good enzymatic activity have been produced. Tang and Breaker isolated self-cleaving RNAs by in vitro selection of RNAs originating from random-sequence RNAs. Some of the synthetic ribozymes that were produced had novel structures, while some were similar to the naturally occurring hammerhead ribozyme. The techniques used to discover artificial ribozymes involve Darwinian evolution.

This approach takes advantage of RNA's dual nature as both a catalyst and an informational polymer, making it easy for an investigator to produce vast populations of RNA catalysts using polymerase enzymes. The ribozymes are mutated by reverse transcribing them with reverse transcriptase into various cDNA and amplified with mutagenic PCR. The selection parameters in these experiments often differ. One approach for selecting a ligase ribozyme involves using biotin tags, which are covalently linked to the substrate. If a molecule possesses the desired ligase activity, a streptavidin matrix can be used to recover the active molecules. Lincoln and Joyce developed an RNA enzyme system capable of self replication in about an hour. By utilizing molecular competition (*in vitro* evolution) of a candidate enzyme mixture, a pair of RNA enzymes emerged, in which each synthesizes the other from synthetic oligonucleotides, with no protein present.

Applications

A type of synthetic ribozyme directed against HIV RNA called gene shears has been developed and has entered clinical testing for HIV infection.

Editing

In some instances, an mRNA will be edited, changing the nucleotide composition of that mRNA. An example in humans is the apolipoprotein B mRNA, which is edited in some tissues, but not others. The editing creates an early stop codon, which upon translation, produces a shorter protein.

Polyadenylation

Polyadenylation is the addition of a poly(A) tail to an RNA molecule. The poly(A) tail consists of multiple adenosine monophosphates; in other words, it is a stretch of RNA which only has adenine bases. In eukaryotes, polyadenylation is part of the process that produces mature messenger RNA (mRNA) for translation. It therefore forms part of the larger process of gene expression. The process of polyadenylation begins as the transcription of a gene finishes. The 3'-most segment of the newly-made RNA is first cleaved off by a set of proteins; these proteins then synthesize the poly(A) tail at the RNA's 3' end. In some genes these proteins may add a poly(A) tail at any one of several possible sites, polyadenylation can therefore produce more than one transcript from a single gene, similar to alternative splicing.

The poly(A) tail is important for the nuclear export, translation and stability of mRNA. The tail is shortened over time and when it is short enough, the mRNA is enzymatically degraded. However, in a few cell types, mRNAs with short poly(A) tails are stored for later activation by re-polyadenylation in the cytosol. In contrast, when polyadenylation occurs in bacteria, it promotes RNA degradation. This is also sometimes the case for eukaryotic non-coding RNAs. The wide distribution of polyadenylation among living organisms indicates that this process evolved early in the history of life on Earth.

Primer on RNA

RNAs are a type of large biological molecules, whose individual building blocks are called nucleotides. The name *poly(A) tail* (for polyadenylic acid tail) reflects the way RNA nucleotides are abbreviated, with a letter for the base the nucleotide contains (A for adenine, C for cytosine, G for guanine and U for uracil). RNAs are produced (*transcribed*) from a DNA template. By convention, RNA sequences are written in a 5' to 3' direction. The 5' end is the part of the RNA molecule that is transcribed first, and the 3' end is transcribed last. The 3' end is also where the poly(A) tail is found on polyadenylated RNAs. Messenger RNA (mRNA) is RNA that has a coding region that acts as a template for protein synthesis (*translation*). The rest of the mRNA, the *untranslated regions*, tune how active the mRNA is. There are also many RNAs that are not translated, called non-coding RNAs. Like the untranslated regions, many of these non-coding RNAs have regulatory roles.

Nuclear Polyadenylation

Function

In nuclear polyadenylation, a poly(A) tail is added to an RNA at the end of transcription. On mRNAs, the poly(A) tail protects the mRNA molecule from enzymatic degradation in the cytoplasm and aids in transcription termination, export of the mRNA from the nucleus, and translation. Almost all eukaryotic mRNAs are polyadenylated, with the exception of animal replication-dependent histone mRNAs. These are the only mRNAs in eukaryotes that lack a poly(A) tail, ending instead in a stem-loop structure followed by a purine-rich sequence, termed histone downstream element, that directs where the RNA is cut so that the 3' end of the histone mRNA is formed.

Many eukaryotic non-coding RNAs are always polyadenylated at the end of transcription. There are small RNAs where the poly(A) tail is only seen in intermediary forms and not in the mature RNA as the ends are removed during processing, notably microRNAs. But for many long noncoding RNAs – a seemingly large group of regulatory RNAs that for example includes the RNA Xist which mediates X chromosome inactivation – a poly(A) tail is part of the mature RNA.

Mechanism

The polyadenylation machinery in the nucleus of eukaryotes works on products of RNA polymerase II, such as precursor mRNA. Here, a multi-protein complex cleaves the 3'-most part of a newly produced RNA and polyadenylates the end produced by this cleavage. The cleavage is catalysed by the enzyme CPSF and occurs 10–30 nucleotides downstream of its binding site. This site is often the sequence AAUAAA on the RNA, but variants of it exist that bind more weakly to CPSF.

Two other proteins add specificity to the binding to an RNA: CstF and CFI. CstF binds to a GU-rich region further downstream of CPSF's site. CFI recognises a third site on the RNA (a set of UGUAA sequences in mammals) and can recruit CPSF even if the AAUAAA sequence is missing. The polyadenylation signal – the sequence motif recognised by the RNA cleavage complex – varies between groups of eukaryotes. Most human polyadenylation sites contain the AAUAAA sequence, but this sequence is less common in plants and fungi.

The RNA is cleaved right after transcription, as CstF also binds to RNA polymerase II. Cleavage also involves the protein CFII, though it is unknown how. The cleavage site associated with a polyadenylation signal can vary up to some 50 nucleotides. When the RNA is cleaved, polyadenylation starts, catalysed by polyadenylate polymerase. Polyadenylate polymerase builds the poly(A) tail by adding adenosine monophosphate units from adenosine triphosphate to the RNA, cleaving off pyrophosphate. Another protein, PAB2, binds to the new, short poly(A) tail and increases the affinity of polyadenylate polymerase for the RNA.

When the poly(A) tail is approximately 250 nucleotides long the enzyme can no longer bind to CPSF and polyadenylation stops, thus determining the length of the poly(A) tail. CPSF is in contact with RNA polymerase II, allowing it to tell the polymerase to terminate transcription. The polyadenylation machinery is also physically linked to the spliceosome, a complex that removes introns from RNAs.

Downstream Effects

The poly(A) tail acts as the binding site for poly(A)-binding protein. Poly(A)-binding protein promotes export from the nucleus and translation, and inhibits degradation. This protein binds to the poly(A) tail prior to mRNA export from the nucleus and in yeast also recruits poly(A) nuclease, an enzyme that shortens the poly(A) tail and allows the export of the mRNA. Poly(A)-binding protein is exported to the cytoplasm with the RNA. mRNAs that are not exported are degraded by the exosome. Poly(A)-binding protein also can bind to, and thus recruit, several proteins that affect translation, one of these is initiation factor-4G which in turn recruits the 40S ribosomal subunit. However, a poly(A) tail is not required for the translation of all mRNAs.

Deadenylation

In eukaryotic somatic cells, the poly(A) tail of most mRNAs in the cytoplasm gradually get shorter, and mRNAs with shorter poly(A) tail are translated less and degraded sooner. However, it can take many hours before an mRNA is degraded. This deadenylation and degradation process can be accelerated by microRNAs complementary to the 3' untranslated region of an mRNA. In immature egg cells, mRNAs with shortened poly(A) tails are not degraded, but are instead stored without being translated.

They are then activated by cytoplasmic polyadenylation after fertilisation, during egg activation. In animals, poly(A) ribonuclease (PARN) can bind to the 5' cap and remove nucleotides from the poly(A) tail. The level of access to the 5' cap and poly(A) tail is important in controlling how soon the mRNA is degraded. PARN deadenylates less if the RNA is bound by the initiation factors 4E (at the 5' cap) and 4G (at the poly(A) tail), so this is why translation reduces deadenylation. The rate of deadenylation may also be regulated by RNA-binding proteins. Once the poly(A) tail is removed, the decapping complex removes the 5' cap, leading to a degradation of the RNA. Several other enzymes that seem to be involved in deadenylation have been identified in yeast.

Alternative Polyadenylation

Many protein-coding genes have more than one polyadenylation site, so a gene can code for several mRNAs that differ in their 3' end. Since alternative polyadenylation changes the length of the 3' untranslated region, it can change which binding sites for microRNAs the 3' untranslated region contains. MicroRNAs tend to repress translation and promote degradation of the mRNAs they bind to, although there are examples of microRNAs that stabilise transcripts. Alternative polyadenylation can also shorten the coding region, thus making the mRNA code for a different protein, but this is much less common than just shortening the 3' untranslated region. The choice of poly(A) site can be influenced by extracellular stimuli and depends on the expression of the proteins that take part in polyadenylation. For example, the expression of CstF-64, a subunit of cleavage stimulatory factor (CstF), increases in macrophages in response to lipopolysaccharides (a group of bacterial compounds that trigger an immune response). This results in the selection of weak poly(A) sites and thus shorter transcripts. This removes regulatory elements in the 3' untranslated regions of mRNAs for defense-related products like lysozyme and TNF-α. These mRNAs then have longer half-lives and produce more of these proteins. RNA-binding proteins other than those in the polyadenylation machinery can also affect whether a polyadenyation site is used,as can DNA methylation near the polyadenylation signal.

Cytoplasmic Polyadenylation

There is polyadenylation in the cytosol of some animal cell types, namely in the germ line, during early embryogenesis and in post-synaptic sites of nerve cells. This lengthens the poly(A) tail of an mRNA with a shortened poly(A) tail, so that the mRNA will be translated. These shortened poly(A) tails are often less than 20 nucleotides, and are lengthened to around 80–150 nucleotides. In the early mouse embryo, cytoplasmic polyadenylation of maternal RNAs from the egg cell allows the cell to survive and grow even though transcription does not start until the middle of the 2-cell stage (4-cell stage in human). In the brain, cytoplasmic polyadenylation is active during learning and could play a role in long-term potentiation, which is the strengthening of the signal transmission from a nerve cell to another in response to nerve impulses and is important for learning and memory formation.

Cytoplasmic polyadenylation requires the RNA-binding proteins CPSF and CPEB, and can involve other RNA-binding proteins like Pumilio. Depending on the cell type, the polymerase can be the same type of polyadenylate polymerase (PAP) that is used in the nuclear process, or the cytoplasmic polymerase GLD-2.

Tagging for Degradation in Eukaryotes

For many non-coding RNAs, including tRNA, rRNA, snRNA and snoRNA, polyadenylation is a way of marking the RNA for degradation, in at least yeast. This polyadenylation is done in the nucleus by the TRAMP complex, which adds a tail that is around 40 nucleotides long to the 3' end. The RNA is then degraded by the exosome. Poly(A) tails have also been found on human rRNA fragments, both the form of homopolymeric (A only) and heterpolymeric (mostly A) tails.

In Prokaryotes and Organelles

In many bacteria, both mRNAs and non-coding RNAs can be polyadenylated. This poly(A) tail promotes degradation by the degradosome, which contains two RNA-degrading enzymes: polynucleotide phosphorylase and RNase E. Polynucleotide phosphorylase binds to the 3' end of RNAs and the 3' extension provided by the poly(A) tail allows it to bind to the RNAs whose secondary structure would otherwise block the 3' end. Successive rounds of polyadenylation and degradation of the 3' end by polynucleotide phosphorylase allows the degradosome to overcome these secondary structures. The poly(A) tail can also recruit RNases that cut the RNA in two.

These bacterial poly(A) tails are about 30 nucleotides long. In as different groups as animals and trypanosomes, the mitochondria contain both stabilising and destabilising poly(A) tails. Destabilising polyadenylation targets both mRNA and noncoding RNAs. The poly(A) tails are 43 nucleotides long on average. The stabilising ones start at the stop codon, and without them the stop codon (UAA) is not complete as the genome only encodes the U or UA part. Plant mitochondria only have destabilising polyadenylation, and yeast mitochondria have no polyadenylation at all.

While many bacteria and mitochondria have polyadenylate polymerases, they also have another type of polyadenylation, performed by polynucleotide phosphorylase itself. This enzyme is found in bacteria, mitochondria, plastids and as a constituent of the archeal exosome (in those archaea that have an exosome). It can synthesise a 3' extension where the vast majority of the bases are adenines. Like in bacteria, polyadenylation by polynucleotide phosphorylase promotes degradation of the RNA in plastids and likely also archaea.

Evolution

Although polyadenylation is seen in almost all organisms, it is not universal. However, the wide distribution of this modification and the fact that it is present in organisms from all three domains of life implies that the last universal common ancestor of all living organisms probably had some form of polyadenylation system. A few organisms do not polyadenylate mRNA, which implies that they have lost their polyadenylation machineries during evolution. Although no examples of eukaryotes that lack polyadenylation are known, mRNAs from the bacterium *Mycoplasma gallisepticum* and the salt-tolerant archaean *Haloferax volcanii* lack this modification.

The most ancient polyadenylating enzyme is polynucleotide phosphorylase. This enzyme is part of both the bacterial degradosome and archaeal exosome, two closely related complexes that recycle RNA into nucleotides. This enzyme degrades RNA by attacking the bond between the 3'-most nucleotides with a phosphate, breaking off a diphosphate nucleotide. This reaction is reversible, and so the enzyme can also extend RNA with more nucleotides. The heteropolymeric tail added by polynucleotide phosphorylase is very rich in adenine. The choice of adenine is most likely the result of higher ADP concentrations than other nucleotides as a result of using ATP as an energy currency, making it more likely to be incorporated in this tail in early lifeforms. It has been suggested that the involvement of adenine-rich tails in RNA degradation prompted the later evolution of polyadenylate polymerases (the enzymes that produce poly(A) tails with no other nucleotides in them).

Polyadenylate polymerases are not as ancient. They have separately evolved in both bacteria and eukaryotes from CCA-adding enzyme, which is the enzyme that completes the 3' ends of tRNAs. Its catalytic domain is homologous to that of other polymerases. Likely, the horizontal transfer of bacterial CCA-adding enzyme to eukaryotes allowed the archaeal-like CCA-adding enzyme to switch function to a poly(A) polymerase. Some lineages, like archaea and cyanobacteria, never evolved a polyadenylate polymerase.

History

Polyadenylation was first identified in 1960 as an enzymatic activity in extracts made from cell nuclei that could polymerise ATP, but not ADP, into polyadenine. Although identified in many types of cells, this activity had no known function until

1971, when poly(A) sequences were found in mRNAs. Initially, the only function of these sequences was thought to be protection of the 3' end of the RNA from nucleases, but later the specific roles of polyadenylation in nuclear export and translation were identified.The polymerases responsible for polyadenylation were first purified and characterized in the 1960s and 1970s, but the large number of accessory proteins that control this process were only discovered in the early 1990s.

Polyadenylation is the covalent linkage of a polyadenylyl moiety to a messenger RNA molecule. In eukaryotic organisms, most messenger RNA (mRNA) molecules are polyadenylated at the 3' end. The poly(A) tail and the protein bound to it aid in protecting mRNA from degradation by exonucleases.

Polyadenylation is also important for transcription termination, export of the mRNA from the nucleus, and translation. mRNA can also be polyadenylated in prokaryotic organisms, where poly(A) tails act to facilitate, rather than impede, exonucleolytic degradation. Polyadenylation occurs during and immediately after transcription of DNA into RNA.

After transcription has been terminated, the mRNA chain is cleaved through the action of an endonuclease complex associated with RNA polymerase.

After the mRNA has been cleaved, around 250 adenosine residues are added to the free 3' end at the cleavage site. This reaction is catalyzed by polyadenylate polymerase. Just as in alternative splicing, there can be more than one polyadenylation variant of a mRNA.

Transport

Another difference between eukaryotes and prokaryotes is mRNA transport. Because eukaryotic transcription and translation is compartmentally separated, eukaryotic mRNAs must be exported from the nucleus to the cytoplasm. Mature mRNAs are recognized by their processed modifications and then exported through the nuclear pore. In neurons mRNA must be transported from the soma to the dendrites where local translation occurs in response to external stimuli. Many messages are marked with so-called "zip codes" which targets their transport to a specific location.

Translation

Because prokaryotic mRNA does not need to be processed or transported, translation by the ribosome can begin immediately after the end of transcription. Therefore, it can be said that prokaryotic translation is *coupled* to transcription and occurs *co-transcriptionally*. Eukaryotic mRNA that has been processed and transported to the cytoplasm (i.e. mature mRNA) can then be translated by the ribosome. Translation may occur at ribosomes free-floating in the cytoplasm, or directed to the endoplasmic reticulum by the signal recognition particle. Therefore, unlike in prokaryotes, eukaryotic translation *is not* directly coupled to transcription.

Structure

5' cap: The 5' cap is a specially altered nucleotide on the 5' end of precursor messenger RNA and some other primary RNA transcripts as found in eukaryotes. The process of 5' capping is vital to creating mature messenger RNA, which is then able to undergo translation. Capping ensures the messenger RNA's stability while it undergoes translation in the process of protein synthesis, and is a highly regulated process that occurs in the cell nucleus.

5' Cap Structure: The 5' cap is found on the 5' end of an mRNA molecule and consists of a guanine nucleotide connected to the mRNA via an unusual 5' to 5' triphosphate linkage. This guanosine is methylated on the 7 position directly after capping *in vivo* by a methyl transferase. It is referred to as a 7-methylguanosine cap, abbreviated m^7G. Further modifications include the possible methylation of the 2' hydroxy-groups of the first 2 ribose sugars of the 5' end of the mRNA. The methylation of both 2' hydroxy-groups is shown on the diagram. The 5' cap looks like the 3' end of an RNA molecule (the 5' carbon of the cap ribose is bonded, and the 3' unbonded). This provides significant resistance to 5' exonucleases.

Capping Process

The starting point is the unaltered 5' end of an RNA molecule. This features a final nucleotide followed by three phosphate groups attached to the 5' carbon.

1. One of the terminal phosphate groups is removed (by RNA terminal phosphatase), leaving two terminal phosphates.
2. GTP is added to the terminal phosphates (by a guanylyl transferase), losing two phosphate groups (from the GTP) in the process. This results in the 5' to 5' triphosphate linkage.
3. The 7-nitrogen of guanine is methylated (by a methyl transferase).
4. Other methyltransferases are optionally used to carry out methylation of 5' proximal nucleotides.

5' Capping Targeting

The Capping Enzyme Complex (CEC) required for capping is found bound to the RNA polymerase II before transcription starts. As soon as the 5' end of the new transcript emerges the enzymes transfer to it and begin the capping process (this is a similar kind of mechanism to ensure capping as for polyadenylation). The enzymes for capping can only bind to RNA polymerase II ensuring specificity to only these transcripts, which are almost entirely mRNA.

5' Cap Function

The 5' cap has 4 main functions:

1. Regulation of nuclear export.
2. Prevention of degradation by exonucleases.
3. Promotion of translation.
4. Promotion of 5' proximal intron excision.

Nuclear export of RNA is regulated by the Cap binding complex (CBC), which binds exclusively to capped RNA. The CBC is then recognized by the nuclear pore complex and exported. Once in the cytoplasm after the pioneer round of translation, the CBC is replaced by the translation factors eIF-4E and eIF-4G. This complex is then recognized by other translation initiation machinery including the ribosome.

Cap prevents 5' degradation in two ways. First, degradation of the mRNA by 5' exonucleases is prevented (as mentioned above) by functionally looking like a 3' end. Second, the CBC complex and the eIF-4E/eIF-4G block the access of decapping enzymes to the cap. This increases the half-life of the mRNA, essential in eukaryotes as the export process takes significant time.

Decapping of an mRNA is catalyzed by the decapping complex made up of at least Dcp1 and Dcp2, which must compete with eIF-4E to bind the cap. Thus the 5' cap is a marker of an actively translating mRNA and is used by cells to regulate mRNA half-lives in response to new stimuli. Undesirable mRNAs are sent to P-bodies for temporary storage or decapping, the details of which are still being resolved. The mechanism of 5' proximal intron excision promotion is not well understood, but the 5' cap appears to loop around and interact with the spliceosome in the splicing process, promoting intron excision.

Chapter 5

Nano Structures

Reversible Architectures

One concern about this approach is the need to use reversible computer architectures. Such architectures are entirely feasible. A wide variety of computations can be done in a logically reversible manner. Bennett concluded: "...for any e > 0, ordinary multitape Turing machines using time T and space S can be simulated by reversible ones using time O(T^(1+e)) and space O(S log T) or in linear time and space O(S T^e)." This general result shows that even arbitrary irreversible computations can be mapped into reversible computations.

Specific irreversible computations can often be mapped into reversible computations with little or no loss of efficiency. Even if we do not adopt new reversible computer architectures, simple applications of reversible computation could be made within the framework of existing architectures.

A typical computer executes a sequence of instructions, and each instruction will typically change the contents of a single register or memory location. Although loading the result of the instruction into a register will normally be irreversible (it erases the previous contents of the register) it is still the case that all other operations performed by the computer during instruction execution could in principle be made reversible.

Thus, although we would have to dissipate roughly kT energy for each bit in the output register for each instruction execution, we need not use dissipative logic throughout the computer (as is currently done).We can go one step further without making significant changes in computer architecture: the simple register-register add instruction $R_1 = R_1 + R_2$ is logically reversible.

With proper hardware, this particular instruction could in principle be made to dissipate as little energy as we desired. While non-reversible instructions (*e.g.*, R1 = 0) would still dissipate greater energy, the overall energy dissipation of the computer could be reduced significantly if a reversible approach were developed and used to implement the reversible instructions.

Of course, once the energy-wasting instructions were identified, compilers would learn to avoid using them. Today, compilers can optimize for speed of computation and for the amount of memory used.

In the future, compilers could also optimize a programme to minimize the energy dissipated during computation, *e.g.*, to minimize the number of bits that are erased.

This would provide an entirely evolutionary path from the current irreversible designs to computer architectures that were as reversible as was practically feasible.

While it is perhaps not obvious how far this trend can go it is clear that a very large percentage of computer operations can be made reversible-perhaps a remarkably large percentage.

Reversible Devices

Several reversible devices have been proposed over the years. Von Neumann and independently Goto proposed the parametron which encodes information in the phase of an oscillation.

Fredkin and Toffoli proposed an electronic reversible logic family which used switches, capacitors and inductors. By turning the switches on and off at just the right times, charge could be transferred between the capacitors through the inductors with minimal energy loss.

By increasing the size of the inductors it would be possible to create a family of ever slower but ever more energy efficient circuits. As the LC time constant is increased, the energy dissipation can be decreased to an arbitrary extent.

A proposal by Likharev based on Josephson junctions "...is particularly significant, because it is a genuine example... of a system that has frictional forces proportional to velocity" according to Landauer.

The parametric quantron is based on Josephson junctions and operates at low temperatures. It should dissipate less than kT when operating at 4 kelvins and with a switching time of 1 nanosecond. Modern "high temperature" superconductors might allow operation at relatively high temperatures, including that of liquid nitrogen.

Even these temperatures would still be a significant disadvantage. Molecular-scale mechanical reversible "rod logic" has been proposed by Drexler which should dissipate less than kT when operating at room temperature at a speed of 50 to 100 nanoseconds. Drexler's proposal is also significantly smaller than the parametric quantron. While the ability to fabricate Drexler's proposal is some decades away, it provides a good argument that molecular-scale reversible devices should be feasible.

Although the parametric quantron can be used in a reversible fashion to achieve low energy dissipation it uses magnetic fields to couple logic devices, requires low temperatures to operate and will likely prove resistant to scaling to the molecular size range. By contrast, Drexler's mechanical proposal might well prove smaller than even the smallest of future electronic devices. Mechanical devices depend fundamentally on the position of the nuclei of the atoms of which they are made.

The position of the nucleus of an atom can be much more precisely known than the position of an electron: quantum uncertainty in the position of the electron is much greater because the mass of the electron is much smaller.

If device function depends on the position of an electron then device size will be larger than if device function depends on the position of the nucleus. Such molecular mechanical proposals will, however, almost certainly be slower because of the greater mass of the nuclei of atoms as compared with the mass of electrons. In this paper we discuss simple methods of using relatively conventional electronic switching devices in a reversible fashion which can be scaled to a very small size. The methods described here allow a simple voltage-controlled switch to be used in a reversible manner, both to implement combinational logic and also to iterate an arbitrary reversible logic function.

Other methods are feasible. For example, helical logic—an elegant form of reversible electronic logic—has recently been proposed by Merkle and Drexler. Voltage controlled switches of various types have long been used in electronic devices. Not all voltage controlled switches are suitable for use in low energy designs because of their intrinsic energy dissipation.

Traditional relays actuated by electromagnets dissipate comparatively large amounts of energy. Even if a switch were suited for low energy operation it would still dissipate energy if it were turned on when there was a voltage across it.

The abrupt rush of current that normally occurs when a switch is turned on would insure this. Most previous logic designs based on switches operated in this manner (though the proposal by Fredkin and Toffoli avoids this problem, and the proposal by Seitz et. al. eliminates dissipative switch closures for combinational logic, although not for sequential logic).

The switch is turned on for the purpose of draining charge when the charge is no longer wanted. That is, a logic "1" might be encoded by a charge on a capacitor.

Turning on a switch to drain away that charge would dissipate energy even if the switch itself were perfectly non-dissipative. A related but distinct line of research also emphasizes the importance of energy dissipation, but assumes that the losses involved in dissipatively switching a wire from one logic state to another are fixed and cannot be altered, *e.g.*, that slowing the switching speed will not reduce energy dissipation.

Even assuming that this is true, clever design of the circuit can greatly reduce the number of times it is necessary to take such a dissipative action. Further research which combines both this approach and the approaches described here would seem fruitful. Even relatively recently some authors have argued that switching devices must fundamentally dissipate more than kT regardless of how slowly they operate. It is worth noting, therefore, that charging and discharging a condenser need not dissipate the energy involved.

For example, a simple LC oscillator will repeatedly charge and discharge a condenser. Although the energy stored on the condenser is $1/2\ CV^2$, this energy need not be dissipated during a charge/discharge cycle.

If the Q of the circuit is high the energy dissipation per cycle will be small. In addition, the energy dissipation of such an oscillator depends on frequency.

By using larger condensers and inductors the frequency of operation can be slowed and the fraction of energy dissipated per cycle reduced. If all linear dimensions of a device are scaled in size in proportion to some characteristic length L then capacitance and inductance are proportional to L while frequency and resistance are proportional to 1/L.

The Q of the circuit is proportional to L so the percentage of the total energy in the oscillator that is dissipated per cycle can be reduced simply by increasing all linear dimensions equally.

Different scaling methods can produce better results. If we assume that a fixed capacitive load is charged through a fixed resistance by a voltage that is ramped up over a period of time T, then the energy dissipated by the resistance is approximately $1/2\ CV^2 \times 2RC/T$ (for $RC \ll T$).

This dissipation can be reduced as much as we might want by increasing T (and leaving the circuit components unchanged). Koller and Athas have called this adiabatic charging. We will call it "thermodynamically reversible" or simply "reversible," relying on context to make it clear that we are referring to an asymptotically non-dissipative process. Note that the term "reversible" can also be used to mean "logically reversible" with no implication that it is asymptotically non-dissipative.

A logically reversible process might be highly dissipative, so it is sometimes necessary to distinguish between "logically reversible" and "thermodynamically reversible. The situation is quite different for a step function in which the voltage is ramped up instantly (T=0). This process dissipates 1/2 CV^2. Charge falls through the fixed potential drop in a dissipative fashion much as water goes over Niagra falls.

Although this is the method in common use today to charge and discharge the various capacitive loads in computer circuits, it is not the method that we will be considering.

By using reversible charging, further increases in the switching time can be used to decrease the energy dissipation per logic operation. This, of course, results in a slower speed of operation leading to a trade-off between speed and energy dissipation. How to arrange matters so that current flows smoothly and reversibly is the principle topic of the rest of the paper.

This is challenging because the basic switching devices are:

- Abrupt and discontinuous and
- Pass through a dissipative intermediate state (which is neither fully on nor fully off) while switching.

The First Method

The first method uses a conventional voltage controlled switch but in a somewhat unusual manner. In essence we never turn a switch on or off when there is either a voltage across it or current going through it.

This mode of operation is reversible. We extend this approach to implement a sequential circuit which can iterate the computation of a reversible logic function. It is also related to "dry switching," a method for using relays which avoids turning relays on or off when there is a voltage across them.

The primary purpose of dry switching is to avoid deterioration of the relay contacts, rather than minimize switching energy.

We also adopt the following convention for turning a switch on and off: when the control input to the switch is at logic "0" (low) the switch is turned on and current can flow from the input to the output or from the output to the input. When the control input is "1" (high) the switch is turned off and current cannot flow. The reader should note that this convention is not the usual one. We adopt it instead of its opposite —where a "1" turns the switch on and a "0" turns it off— because it is more intuitive when applied to the second proposal discussed later in this paper, the CCD-based reversible logic element.

The proposed mode of operation is fundamentally clocked and multi-staged. The outputs of stage i drive the inputs of stage i+1. For the moment, we assume the output of stage 0 is simply given. That is, a set of input lines are assumed to have fixed voltages upon them representing the input data.

The value of the succesive logic levels can then be computed in turn. The logic values (voltages) produced by stage 1 are computed from the logic values given by stage 0; The logic values produced by stage 2 are computed from the logic values produced by stage 1; etc. Simple pass logic can be employed to implement combinational logic. A "NOR" operation in this logic is simply two switches in series.

A clock signal (initially at logic 0) is connected to the input of the first switch. If the control inputs of both switches have settled to logic 0, then both switches will conduct. The clock input connected to the first switch is slowly raised (from logic 0 to logic 1). Both switches will conduct and the output will gradually shift to a logic 1. If either input is 1, one of the switches will not conduct and the output will remain at 0.

Making an N-level Circuit

We assume that we have an N level combinational circuit, and that N "clock signals" are provided. Initially, clocks 1 through N are low (logic 0), and the outputs of stage 0 are at some preset values. The outputs from stage 1 are computed as simple logical functions from the outputs of stage 0. The outputs from stage i are used as the inputs to stage i+1. A "NOR" and "NAND" gate from stage 1. The NAND gate is logically complete so any arbitrary boolean function can be implemented by using an appropriately connected collection of them. The clock signals are now raised in turn.

The output of "stage 0" is assumed to be valid prior to the start of operations, and clock 0 is unspecified (for now). Clock 1 is raised, and the outputs from stage 1 become valid. Clock

2 is raised, and the outputs from stage 2 become valid. This sequence continues, until clock N is raised, and the outputs from stage N become valid. A single stage in the N level circuit could in principle implement an arbitrary boolean switching function by an appropriate arrangement of switches.

This might require an exponential number of switches and so multiple stages appear to be useful. Also, although for simplicity we have assumed that the outputs of stage i are used directly only in stage i+1, they could equally be used in stages i+2, i+3, etc. We will not consider these additional complexities here because they are secondary to the fundamental issues.

Latches

What has been described so far will allow the construction of an N level reversible combinational logic circuit. To be useful, we must also specify a latch which can store the output of the circuit. This can be done by holding charge on a capacitor. The outputs of stage N are connected through switches to a latch stage, stage N+1.

The latch requires its own clock. Initially, the latch clock is 0 and allows charge to move freely into the capacitor. When the output from stage N is valid, the latch clock goes high (changes to logic 1) and prevents charge from escaping.

Clock N can now go low (to logic 0) which will make the output from stage N also go low, but the output of the latch will not be effected by changes in the output from stage N because it is cut off. Of course, having once "locked up" the outputs in the latches, we cannot release the information in the latches.

If we should again set the clock latch low, allowing charge to flow off the capacitor, then we would be turning on a switch when there was a voltage across it. This is forbidden, and so the problem of removing information from a latch is non-trivial.

Unwriting

Rather than simply "erasing" the information on the latch by letting the latch clock go low, we need to "unwrite" the information. Unwriting differs fundamentally from erasing.

When we erase, we do not know the state of the latch, we simply turn on the switch holding the charge on the latch and let the charge flow away. Whether the latch held a "0" or a "1" is immaterial.

In the process, we must necessarily dissipate energy because erasing information is irreversible. By contrast, when we "unwrite" a latch we actually have an input signal that tells us the state of the latch. We can set the latch clock low and connect the input signal and the value stored on the latch.

Because (by definition) the input signal is in the same logical state as the latch, there is no voltage across the switch when we turn it on. We can now let the input signal go to 0 slowly, thus draining charge from the latch.We can indefinitely reduce the energy dissipated by this process by slowing the speed with which we drain charge from the latch.

Summary of Combinational Logic

In our earlier description, we left the source of valid input a mystery. The outputs of stage 0 held valid data, and how this happy state of affairs came about was unspecified. Now we can rectify this omission: stage 0 is a latch. This implies that clock 0 must be kept high to prevent the charges in stage 0 from leaking away. Further, stage N+1 is a latch, and holds the output. We still leave the mechanism that initially put proper charge levels into the capacitors of stage 0 a mystery, but it is now clear that stage 0 will continue to hold a valid signal until clock 0 is lowered and the charge is allowed to escape.

The latches of stage 0 will also be called the input latches or simply the input, while the latches of stage N+1 will also be called the output latches or simply the output.

We can now specify our initial state as one in which appropriate charges are held on the input, clock 0 is high (to prevent charge from the input latches from leaking away), while clocks 1 through N are held low; as is clock signal N+1 (that is, there are no charges held on the output latches and the switch connecting the output latches to the final logic stage is turned on).

We raise clock signals 1 through N in turn, computing the values of each logic level in turn. Finally, we raise clock signal

N+1, thus trapping the output signals in the output latches. We are then free to lower clock signal N, then N–1, then N–2, etc. until we have finally worked our way back to the first logic level. When the computation is done, all clock signals from 1 through N are held low, while clock signals 0 and N+1 remain high. Charge is trapped in stage 0 corresponding to the input, and charge is trapped in stage N+1 corresponding to the output. The outputs of stage 0 and N+1 are valid. All other outputs are 0.

Iteration of a Reversible Logic Function

At a higher level of abstraction, we can say that the outputs of stage N+1 are equal to some combinational function F applied to the outputs of stage 0, or output = F(input). Of course, we're now stuck. We cannot unwrite the input latches because we don't know what information is stored in them. We can solve this dilemma by requiring that F be reversible. That is, we demand that F^–1(F(input)) = input for all possible values of the input on stage 0. This allows us to compute the values stored in stage 0 from the values stored in stage N+1. We need merely compute F^–1(output).

That is, we add additional logic stages N+2, N+3..., N+M which compute the inverse of F. The output of stage N+M will then be the same as the input. Because we now have data values which are identical to the contents of stage 0, we can unwrite stage 0 in a fully reversible fashion.

That is, we can lower clock 0 and allow the charge on the capacitors in stage 0 to be connected with the output signals from stage N+M. Because they are identical the energy loss from this operation can be made arbitrarily small.

Then, clock signal N+M can be lowered, clock signal N+M–1 can be lowered, etc. until we finally lower clock signal N+2. Clock signal N+1 (the latch control clock signal for the output) is high (to hold the output steady in the output latches) and remains high.

To review the cycle of operations from the initial state: we start with a valid input in stage 0, with clock signal 0 held high (to hold the charge in the latches of stage 0) and with all other clock signals held low. From this intial state, we start the computation by raising clock signals 1 through N in turn, computing F(input). Clock signal N+1 is then raised to hold the output values in the output latches of stage N+1, and then clock signals N, N–1, N–2..., 1 are lowered in their turn to "uncompute" the calculation of F. We then compute F^-1(output) by raising clock signals N+2, N+3, N+4..., N+M. This computation produces exactly the values that are stored in stage 0, and so we can now unwrite the contents of stage 0 by lowering clock 0: this turns on the switches connecting the latches of stage 0 to the output of stage N+M.

We then uncompute the calculation of F^{-1}(output) by succesively lowering clock signals N+M, N+M–1, N+M–2..., and N+2. Finally, we again raise clock 0 to again isolate the input (which is now all 0). At the end of this process the outputs are stored in stage N+1 and the input is all 0. The entire process was carried out in a fully reversible manner, and so energy dissipation can in principle be made arbitrarily small.

The only thing we need finally do is exchange the data in the input and output latches (we assume that the input and output latches have the same number of bits). To do this, we first transfer the data in the output latches into the input latches and then "unwrite" the output latches. While the circuitry for this is not entirely trivial, it should be clear that the principles developed should allow us to do this. The sequence of steps involved is gone over in detail in the following several paragraphs.

To exchange the data between the input latches and the output latches, we require four additional sets of switches and four additional sets of clock lines. We illustrate the process by considering the exchange between a single input latch and a single output latch. Initially, EC_1 and EC_2 are set to 0 while EC_3 and EC_4 are set to 1. The switches S_3 and S_4 are turned off. S_1 is turned on, because the input latch holds a 0. S_2 could be either on or off, depending on the data on the output latch.

We first transfer data from the output latch to the input latch. EC_3 is lowered turning on S_3. Because the input latch holds 0 and EC_2 is set to 0, turning on S_3 is permitted. EC_2 is then raised to 1. If S_2 is on, charge is transferred to the input latch. If S_2 is off, charge is not transferred. Thus, if the output latch holds a 1 the input latch will be set to 0. If the output latch holds a 0, the input latch will be set to 1. In the process of transferring the data from the output latch to the input latch, we have also inverted it. To correct this omission would require an inverter, or some modifications to F and F^{-1}. This would take an additional switch and an additional clock line, as well as additional explanation. However, would rapidly become cluttered so we have chosen to omit the inverter to clarify the exposition. This ommision has no impact on the basic concept.

Having set the input latch with data from the output latch, we now raise EC_3 and turn off S_3. Finally, we lower EC_2. This completes the transfer of data from the output latches to the input latches. We must now unwrite the output latches. Unwriting the output latches is similar to setting the input latches, but the exact sequence of operations is changed. Our first step is to raise EC_1. We then lower EC_4 turning on S_4. EC_1 is high, so if there is charge on the output latch then we have turned on a switch when there is no voltage across it, which is permitted.

If there is no charge on the output latch, then there will be charge on the input latch (because we transferred the complement of the output latch to the input latch in the earlier steps) and this charge will turn off S_1. Therefore, the output latch and EC_1 will not be connected. Again, we have not turned on a switch with a voltage across it.

At this point, if the output latch is charged it is connected to EC_1. By lowering EC_1 we can drain the charge from the output latch. We then raise EC_4 turning off S_4. This isolates the output latch again and completes the unwriting of the output latch from the input latch.

The clocking sequence, in brief, is: lower EC_3, raise EC_2, raise EC_3, lower EC_2, raise EC_1, lower EC_4, lower EC_1, raise EC_4. This sequence of operations has exchanged the data on the input latch and the output latch, and did not require turning on a switch when there was a voltage across it. The careful reader might wonder about the charge that might be trapped on the segment between S_1 and S_4 or between S_2 and

S_3. These segments are initially 0, and are restored to 0 upon completion of all operations.

During the actual transfer and unwriting steps, the charge on these segments is always known and does not result in turning on a switch with a voltage across it. An explicit capacitance attached to this segment would not alter the conclusions reached here for this specific design.

In other designs it might be useful to eliminate such segments. Whether or not such segments can be eliminated depends on the specific implementation. Elimination can often be done by appropriate physical design of the switches, effectively resulting in a single switch with two control inputs.

We have now completed the full cycle of operations required to compute F(input). We can go on to compute F(F(input)), F(F(F(input))), etc. We assume F is computationally useful (*e.g.*, it implements a single instruction or step of a reversible computation). As discussed earlier, reversible functions can be used to carry out arbitrary computations.

Problems with Inaccurate Voltage

While the system described is reversible in principle, the energy dissipation in a practical implementation would depend upon the accuracy of the voltage. When a switch is turned on, the logical state of the input and output values is the same, but the physical state might differ.

A small difference in the voltages at the input and output of the switch would result in energy loss as the switch was turned on. Even with this restriction, it should be possible to achieve significant reductions in energy dissipation. The energy dissipated by dissipatively discharging a capacitive load is,

$$\text{Energy} = 1/2\ CV^2$$

where C is the capacitance and V is the voltage. In conventional (irreversible) logic systems, the voltage V would be the voltage of the power supply. In a reversible system of the type described here the voltage V would be the error in the voltage of the power supply, rather than the voltage of the power supply itself.

Thus, by using reversible instead of irreversible logic, we have effectively reduced the value of V. With a power supply accurate to 5%, the energy losses from charging and discharging the load capacitances associated with the electrical connections would be decreased by a factor of 20^2 or 400. That is, energy losses would be a function of the error in the voltage of the power supply, rather than a direct function of the voltage of the power supply itself.

There does not appear to be any reason in principle why the error in the voltage of the power supply could not be made as small as desired. A more fundamental reason for long term optimism is the observation that electric charge is quantized. In future systems in which the number of electrons that represents a logic “0” or “1” is both finite and small, it should be possible to reduce the energy loss caused by slight inaccuracies in the voltage supply to 0.

It is already possible to control the passage of individual electrons in a circuit. Such highly precise control should eventually make a fully reversible electronic switch with extremely small energy losses feasible.

Problems with FETS: Voltage Degradation

Switching elements that do not degrade the voltage of the signal being switched are feasible in principle, as shown by the electrostatic relay. CMOS transmission gates can also be used as switches without concern about voltage degradation. Transmission gates require that both the control signal and its inverse be provided. A simple method of satisfying this requirement would be to use two-rail logic. To use nMOS (or pMOS) by itself, we must overcome certain shortcomings.

In particular an nMOS switch, when used in pass logic, will cause voltage degradation. As a consequence, deterioration of the voltage through multiple logic levels would impose a severe limit on the capabilities of the system.

Boosting voltage in a fully reversible fashion is, however, straightforwards in principle. The voltage on a capacitor will increase if the capacitance decreases. Variable capacitors are well known based both on physical motion of the capacitor plates, or on changes in charge distribution secondary to a voltage change. Varactors in which the capacitance C changes by a factor of 10 or more in response to a voltage change of little more than a volt are feasible. The approach outlined here is abstractly similar to the proposal by Joynson et. al. Before starting, the switch is turned on and current can flow onto the varactor. The high-voltage clock then opens the switch, trapping the charge on the varactor.

Finally, the capacitance of the varactor is reduced. Because the voltage on a capacitor is V=Q/C (where V is voltage, Q is charge and C is capacitance) the voltage across the varactor will increase as its capacitance decreases.

Because the voltage on the high-voltage clock is large, there is more than sufficient control voltage to insure that the switch remains turned off even though the voltage across the switch has been increased.

The low-voltage input, boosted to a medium-voltage output, can then be used to switch a low-voltage signal, which can again be used as the input to a reversible voltage booster. To reverse this cycle of operations, the capacitance of the varactor is returned to its original value.

This will also return the voltage on the output to its original value, and the high-voltage switch can be turned on with minimal energy loss. This energy loss will be a function of the square of the error in the voltage regulation. If the number of electrons that charge the varactor (and any stray output capacitance) is reduced to a small integer number, it should eventually be possible to eliminate losses due to small inaccuracies in voltage regulation.

Seitz et. al. commented that "...the isolation transistor can turn on while there is voltage across it, and accordingly, it dissipates power in charging or discharging the bootstrap node.

The goal of exporting all of the dynamic power is elusive." Translating from their language to the language used here, they sometimes allowed the high-voltage clock to turn on the high-voltage switch when a voltage existed across the switch, thus dissipating energy.

They did not consider the computation of F^-1 and "unwriting" operations, and so were sometimes obliged to engage in the dissipative operation of discharging the gate of a FET to ground. In essence, they were able to eliminate unnecessarily dissipative operations for combinational logic, but not for sequential logic. As shown here, appropriate design of a reversible sequential circuit eliminates the unnecessarily dissipative steps.

Problems with FETS

Leakage Current

Another problem with FET's is leakage current. Even a FET in which the gate voltage is sufficient to insure that the FET is completely cut off will still pass a small current. If our objective is to achieve energy dissipations below kT, any significant leakage current is intolerable. Leakage current occurs because thermally generated electron-hole pairs are created in the semiconductor material of the FET.

Thermal agitation will occasionally cause an electron in the valence band (*e.g.*, an electron which normally would not be able to move because it was fixed in place as part of a covalent bond) to leap into the conduction band. An electron in the conduction band is free to move throughout the crystal and so to conduct electric current.

The energy gap between the valence band and the conduction band will govern the concentration of electrons that are available to carry leakage current. In silicon, the bandgap is 1.12 electron volts.

At 300 kelvins, this will produce 1.45×10^{10} electrons per cubic centimeter of silicon. This "intrinsic carrier concentration" will result in dissipative energy losses.

Three obvious methods are available for reducing the leakage current:

- *Reduce the Physical Size of the Device*: An intrinsic carrier concentration of 1.45×10^{10} electrons per cubic centimeter can also be expressed as 1.45×10^{-11} electrons per cubic nanometer. If the active region of the device was smaller than a few millions of cubic nanometers, then the probability of finding an undesired charge carrier in that region would be relatively small.
- *Reduce the Temperature*: *The intrinsic carrier concentration is*:

$$ni = km \exp(-Eg/2kt)$$

where ni is the intrinsic carrier concentration, km is a constant specific to the material, and Eg is the bandgap energy. In the case of silicon, km is about 3.1×10^{20}. At a temperature of 77 Kelvins, ni is 7.4×10^{-17} electrons per cubic centimeter. CCD's with dark currents of a few electrons per pixel per hour have been fabricated.

- Select a material with a larger bandgap. Diamond has a bandgap of 5.47 electron volts, with an intrinsic carrier concentration at room temperature in the range of 10^{-26} electrons per cubic centimeter. Boron provides a good shallow dopant, and p-type diamond has been made. Several shallow dopants for n-type material are plausible. Diamond MOSFETs with boron doping have been fabricated. An implementation could use purely p-type materials, if that proved useful.

  A variety of wide band-gap materials exist allowing a wide range of tradeoffs between energy dissipation and other parameters. It should be feasible to produce acceptable leakage currents using a variety of different materials in a variety of different geometries under a broad range of operating temperatures. The use of small FET's using a wide band-gap material is particularly attractive for room temperature operation. In such a structure the probability of finding even one valence electron in the conduction band is astronomically small.

Thermally generated leakage current is also of concern in many types of varactors. A reverse- biased P-N junction illustrates the problem: electron-hole pairs generated thermally in the charge depletion region are swept out of the region, thus conducting current.

Leakage current in such varactors can in principle be made quite small for the same reasons that apply to leakage current in a FET.

Reversible CCD-Based Logic

An alternative approach, which eliminates the need for high accuracy voltage supplies, is to eliminate the need to turn on the switch when there are charge carriers present. This might be viewed as only turning the switch on and off when the voltage across the switch is 0 volts, but this is not a fully adequate description. Even a 0 volt signal is subject to fluctuation and noise. If, however, we try to put a positive voltage onto the switch (*e.g.*, the inputs and outputs of the switch both have a positive voltage), and if the charge carriers in the switch are electrons, then it is possible to remove all the charge carriers from the channel region.

Any charge carriers will be actively swept out of the channel region by the positive voltages on the input and output leads. Once all the charge carriers have been removed from the channel region, then no more current will flow.

Precise regulation of the positive voltage on the input and output of the switch is not required, for both need merely be "sufficiently positive" to insure removal of all charge carriers from the channel region.

Once the charge carriers have been removed then turning the switch on and off can be done in a fully reversible manner. Small errors in the supply voltage can no longer lead to dissipative current flows when the switch is turned on, because no current can flow.

For purposes of this discussion we will assume our switch is a FET with electrons as the charge carriers and that the electrons are removed from the FET by a positive voltage on the input and output leads. (We could equally well have assumed the charge carriers are holes and that a negative voltage is placed on the input and output leads: for purposes of explanation the choice is arbitrary).

As discussed earlier, it is essential that the device be designed so that the concentration of intrinsic charge carriers is sufficiently low that energy losses from this source can be neglected. Further, the density of device defects must be low enough that defect sites do not appreciably alter device performance.

We will simply assume that the material is free of defects. Reduction of device size coupled with decreases in defect rates should eventually allow construction of devices with a sufficiently low defect rate that energy dissipation will not be significantly degraded. In this approach, 1's and 0's are represented by a packet of electrons or the absence of a packet of electrons, respectively. The electrons are stored in potential wells which we will call "buckets." Each packet of electrons is treated as a unit: we never split or merge packets, nor do we create or destroy packets (except when the system is initialized, which would ideally occur only once).

While all packets have approximately the same number of electrons, it is not critical that each packet have exactly the right number of electrons. Whether a packet has 99, 100, or 101 electrons won't matter.

The use of charge packets is most familiar in the context of CCDs (Charge Coupled Devices), or alternatively in the context of BBD's, or Bucket Brigade Devices. In a CCD, charge packets are transferred from potential well to potential well serially along a row of devices.

The amount of charge in each packet is a measure of some analog quantity. Such devices are primarily used as memory elements or shift registers, not for computation. Proposals to use the same basic concepts for logic operations have been made since the invention of the CCD.

Sometimes called "DCCL," or Digital Charge Coupled Logic, these proposals have explored the advantages of using charge packets as the basic logic element. The primitive logic elements in prior proposals, however, were highly dissipative. For example, Zimmerman et. al. proposed logic operations whose computation required merging packets, discarding packets by draining the charge to ground, etc.

The goal of thermodynamically reversible logic was not an objective. Tompsett proposed circuits in which the potential generated by a charge packet in a CCD potential well was coupled to the gate of a FET. This FET would then be turned on or off depending on whether a charge packet was present or absent in the potential well. This was then used to create another charge packet (rather than to control the movement of an existing charge packet). The creation of a new charge packet would compensate for imperfect charge transfer.

NAND and NOR operations could be performed by applying the potential from two potential wells to two gates of two FETs, which could then be connected in serial or parallel. Again, the objective of thermodynamic reversibility was not considered. The method of sensing charge in the potential well was dissipative and the sensed packets, no longer needed, were to be dissipatively discarded. For simplicity, we will assume that packets never gain or lose electrons (due to defects in the material or to thermal noise).

In fact, present day devices can gain or lose charge at the rate of a few parts per million. As discussed earlier, such gains or losses should eventually be reduced to insignificant levels. Even if we allow packets to gain or lose electrons at some low rate, we could periodically "refresh" the packets, and so prevent malfunction.

Tompsett first proposed methods of refreshing packets (though he did not consider reversible methods). To the extent that the packet has gained or lost an unpredictable amount of charge, "refreshing" the packet must fundamentally dissipate energy. Other than this fundamental loss, however, we wish to keep energy dissipation to a minimum.

A reversible method of refreshing a packet would be to use the old packet to charge a bucket in a reversible fashion (as discussed in the first method), and then to "unwrite" or discharge the bucket holding the old packet by using the new packet as the source of information.

This method of refreshing a packet would, of course, dissipate energy as a function of the errors in the voltages involved (as discussed earlier). If the refresh operations were done infrequently (*e.g.*, a packet would be refreshed only after a large number of logic operations) then the energy dissipation per logic operation from this source would be quite small.

While CCD and DCCL circuits are today implemented on a semiconductor surface using planar technology, for purposes of this discussion we will initially ignore this geometrical constraint and instead focus on more fundamental issues. A planar version of the ideas described here is given in the section on "A Planar Version of RCT Logic." Illustrates the "hydraulic" model of charge transfer in a CCD. Charge is transferred from one potential well to the next much like water would flow to a gradually sinking region in a pond. This process can be thermodynamically reversible when appropriately implemented. While water has a sharply defined surface, the energy of an electron at some non-zero temperature T is statistical.

To conclude that extremely small perturbations in the potential might cause significant changes in the pattern of charge flow, in fact thermal noise will allow electrons to surmount small barriers *e.g.*, where the barriers have a potential similar in magnitude to kT/e or.026 volts at room temperature (where e is the charge on an electron). Small perturbations and errors in the potential do not fundamentally create dissipative problems.

Tranferring Charge with Little Dissipation

Before discussing the more complex operations needed for computation, we first consider a simple sequence that transfers charge from a source bucket to a destination bucket with as little energy loss as desired.

The use of metallic plates or a metallic connection between the two plates would cause unwanted energy dissipation). At first, both the source clock and destination clock are + (positive), and the charge packet is entirely in the source bucket.

The clocking sequence will then shift the charge from the source bucket to the destination bucket in a series of steps, where each step can dissipate an arbitrarily small amount of energy. As we move from state 1 to state 2, the destination clock is changed from + to -. This change results in negligible energy dissipation, because no charge can move at this time.

As we move from state 2 to state 3, the source clock is changed from + to -. The charge carriers in the charge packet (electrons in our example) now move out of the source bucket. Because the destination clock is negative, the electrons will migrate smoothly out of the source bucket. (If the destination clock were + at this point instead of -, the electrons would move out of the source bucket and then "fall down hill" into the destination bucket, dissipating energy).

As we move from state 3 to state 4, the destination clock changes from-to +, thus slowly attracting the electrons and gathering them into the destination bucket. Finally, as we move from state 4 to state 5, the source clock changes from-to + restoring the system to its original condition, but with the charge packet in the destination bucket instead of the source bucket.

The primary source of energy dissipation in this sequence is from state 2 to state 3, and from state 3 to state 4. These transitions involve charge migration, and will therefore result in dissipative losses.

As we slow the clock frequency the clock voltages change more slowly, the charges move more slowly, and hence dissipate less energy. Because energy dissipation is a function of the square of the current, reducing the frequency of operation by a factor of two will also reduce the current by a factor of two but will reduce the energy dissipation by a factor of 4. This will reduce the energy dissipated per logic operation by a factor of 2. Thus, by sufficiently slowing the clock frequency we can dissipate as little energy per operation as might be desired.

It is important to notice that these operations do not erase information. We start with a charge packet in a single potential well on the left, and we end with a charge packet in a single potential well on the right.

During the course of moving the charge packet we slowly change the shape and position of the potential well, but at no point do we merge potential wells. This transition eliminates the right hand potential well without merging it with the well on the left. An attempt to merge the two potential wells would be dissipative. The transition from step 2 to step 3 slowly changes the shape of the potential well, but the charge packet is always contained in a single potential well and is always at or near thermodynamic equilibrium within that well.

The Primitive Operation

We now consider the somewhat more complex operation that is required to support computation in a reversible fashion.

The single primitive operation is: Transfer charge out of a source bucket and into an empty destination bucket if a "condition" bucket holds a "0", but do not transfer the charge if the condition bucket holds a "1". If the transfer does not take place (the condition bucket is "1"), the destination bucket need not be empty. "Empty" in this context means "has no charge carriers present." That is, an "empty" bucket is in logic state "0." We will call this device a 3-bucket Reversible Charge Transfer device, or (where context makes it clear which type of RCT we are referring to) simply an RCT.

The RCT is a reversible DCCL device. At a logical level the RCT is related to (but somewhat different from) a switch gate implemented using charge packets instead of colliding billiard balls. The particular logic function described here is but one example of the range of logic functions that involve the reversible transfer of charge from one bucket to another.

During normal operation, RCT devices:

- Never merge or split charge packets
- Never discharge charge packets to ground, and
- Move charge packets in a thermodynamically reversible fashion from a set of source buckets to a set of destination buckets, conditional on the presence or absence of charge in a set of condition buckets.

In this context thermodynamic reversibility means that energy dissipation per device operation, in the absence of manufacturing defects, could be made much smaller than kT simply by slowing the speed of operation.

The number of charge packets is conserved and therefore RCT devices are conservative logic devices (in conservative logic devices, the number of 1's is conserved during the computation, so the number of 1's at the input is the same as the number of 1's at the output).

We have described the 3-bucket RCT, which has one condition bucket, one source bucket, and one destination bucket. More complex RCT's have more buckets and the pattern of charge transfer is more complex. Some of these RCT's will be considered later.

The simple 3-bucket RCT operation might be written in C as:

- If (Condition==0) {Destination = Source; Source = 0}; while a somewhat briefer notation would be:
- If Condition is 0 then Destination = Source. (Note that in this briefer notation, the "=" sign is used to indicate both the assignment of the Source value to the

Destination, and setting the Source to 0 after the assignment).

The precondition that must be true prior to execution of the 3-bucket RCT is:

(Destination==0) OR (Condition==1)

Violation of this precondition would result in an unwanted dissipative step, which is banned in RCT devices. While it might at first seem that satisyfing this precondition would make the 3- bucket RCT data dependent, it is in fact possible to design a circuit in which we know that the precondition is satisfied without knowing what values are presented as inputs to the circuit. On the left, the legal initial states of the RCT are shown, while on the right the final states (after the sequence of clock signals that cause the device to step through a single cycle of operations) are shown.

It is interesting to note that there is only a single legal, non-trivial change of state: when the input is 1, the condition is 0, and the output is 0. In all other cases either the input and output states are the same or the input state is illegal. The illegal states would cause undesired energy dissipation.

Fredkin Gate with 3-bucket RCT

The 3-bucket RCT is logically complete, for we can implement a Fredkin gate with it. The construction used here is related to (though diffrent from) that used by Fredkin and Toffoli for constructing a billiard ball Fredkin gate from a billiard ball switch gate.

A Fredkin gate has three inputs and three outputs. The inputs are the control input A, and two signal inputs B and C. The outputs are A', B', and C'. If the control input A is 0, then A' = A, B' = B, and C' = C. If the control input A is 1, then A' = A, B' = C, and C' = B. That is, the output of the Fredkin gate is identical to the input if A = 0, but B' and C' are exchanged if A = 1.

The following sequence of simple RCT operations will implement a Fredkin gate: A, B, and C hold arbitrary input values. InitiallyOne is set to a logic "1" (the presence of an electron packet). All other variables are initialized to logical "0" (the absence of an electron packet).

- If A is 0 then C' = C.
- If A is 0 then B' = B.
- If A is 0 then NotA = InitiallyOne.
- If NotA is 0 then C' = B.
- If NotA is 0 then B' = C.
- If AlwaysZero is 0 then A' = A.
- If A' is 0 then InitiallyOne = NotA.

Steps 1 and 2 simply transfer charge from B and C into B' and C' if A is 0. B' and C' are initially empty, so this transfer satisfies the RCT preconditions.

Step 3 computes the logical negation of A and leaves that logical negation in "NotA."

Steps 4 and 5 transfer charge from B and C into C' and B', exchanging the outputs. The precondition is satisfied, for if A is 0, then NotA will be 1 and the transfer will not take place. If the transfer does not take place, the prior contents of B' and C' are irrelevant.

If A is 1, then NotA will be 0 and the transfer will take place. In this case, B' and C' will be empty (have logic values 0) and so the transfer will take place correctly.

Step 6 copies A into A' unconditionally.

Step 7 restores the value of InitiallyOne if it was altered during the computation.

This implementation of a Fredkin gate will take three arbitrary input values and, after a sequence of RCT operations, produce three output values. We could clearly apply N Fredkin gates to 3N input values and produce 3N output values at the same time.

We require that all output values be initially 0. After finishing the sequence of clocking operations, the inputs are all 0 and the outputs hold arbitrary values. At this point, the "input" and "output" can be logically exchanged (*e.g.*, no physical operation takes place, but we relabel the "input" variables as "output," and the "output" variables as "input").

Following this logical exchange, we have reestablished the preconditions for the next sequence of operations: the (just relabeled) inputs again hold arbitrary logic values, while the (just relabeled) outputs hold logic 0's.

We now drive a (different) set of clock lines to apply the (different) Fredkin gates to generate the next "state" of the computation. Viewed globally, this is a "ping-pong" or "double buffer" scheme for computing successive values of the global state of the system. If we call the two sets of variables A and B, and we call the two logic functions for the "next state" of the system F and G, then we first compute B=F(A). B is initially all zeroes, while A holds arbitrary logic values.

On completion of the computation of F, A is all zeroes while B holds arbitrary logic values. We then compute A=G(B). Because G uses B as its input and A as its output, the condition that the outputs be all zeroes is met during the compuation of G. This returns us to the state where A holds arbitrary logic values and B holds all zeroes. We can repeat this cycle indefinitely.

Implementation of a 3-bucket RCT

Many kinds of implementations would be feasible, for example Silicon-On-Insulator technology (the FET used for the transfer switch would be "tilted on its side" in this approach). Note that many of the conductive paths are really semiconductors. Each bucket also has one semiconductor plate. The gate is also a semiconductor. Use of metallic wires for these components would compromise device function. A planar RCT is described in a later section.

This RCT can be viewed as a combination of three CCD potential wells and a FET. A charge packet is first transferred from the "condition" potential well to the gate of the FET. The charge packet in the "source" potential well is then moved to the source of the FET and (the charge on the gate of the FET permitting) through the channel and out the drain of the FET into the "destination" potential well.

Tompsett proposed sensing the presence or absence of a charge in a CCD potential well by connecting the potential well via a doped region connected ohmically to a metalization that became the gate electrode of a FET. Charge injection from the gate into the doped region, while small if the capacitance of the gate is small, still introduces an unwanted source of energy dissipation.

Other aspects of Tompsett's proposal were also irreversible or highly dissipative. In the present proposal, we require that both the gate of the FET and the connecting path from the potential well to the gate be a semiconductor. The only charge carriers in this semiconductor path would be those deliberately introduced as a charge packet.

An alternative approach which might be more convenient in a simple planar implementation would be to introduce a capacitive coupling into the path from the potential well to the gate. The gate could then be metallic, one plate of the capacitor (the plate connected to the gate) could also be metallic, but the other plate and the connection to the potential well could be a semiconductor.

Potential could be coupled through the capacitor, but charge carriers would be prevented from moving into or out of the semiconductive region and hence charge injection from the gate electrode would not be a problem. Initially, a charge of unknown value is present in the source bucket. The destination bucket is either empty (if the condition bucket holds a 0) or has an unknown charge (if the condition bucket holds a 1). Clocks 1, 2, and 3 are all positive. All charge carriers (electrons) have been drawn onto the semiconductor plates of the buckets.

The conventional plates are connected to the clocks. There is a net positive charge on each bucket creating a potential well which holds the electrons in the bucket and prevents them from migrating away. The channel regions of all switches have no charge carriers present. Clock 4 is driven negative with a large voltage. This large voltage cuts off and isolates the three buckets involved from any other buckets to which they might normally be connected. Additional switches controlled by clock 4 might be required to further isolate the three buckets in a larger circuit, but are not shown in this small example.

The two gray channels do not actively participate in the process of reversible charge transfer, they are present merely to show that connections to other circuit components will be present in a real system. Clock 3 is driven negative, forcing charge from the condition bucket onto the gate of the transfer switch. If no charge is present on the condition bucket, no charge will be put onto the transfer gate, while if charge is present then it will be transferred to the transfer gate, cutting off the transfer switch. Note that the circuit is so arranged that only a medium voltage will be developed on the transfer gate, and this medium voltage can be contained by the high voltage of the cut off clock, clock 4.

Clock 2 is driven negative. If there is charge on the destination bucket, then the transfer switch is turned off and the charge on the destination bucket goes nowhere. If there is no charge on the destination bucket, then whether the transfer switch is turned off or on is immaterial, for no charge will be transferred.

Clock 1 is driven negative. If the transfer switch is turned off, no charge will be transferred. If the transfer switch is turned on, charge will move gradually through the transfer switch into the destination bucket. Because clock 2 is negative, charge will not "fall down hill" in a dissipative fashion when moving into the destination bucket.

Clock 1 has a low voltage so a low voltage is generated during the transfer operation. The medium voltage on the transfer gate is sufficient to block charge transfer if charge transfer is not supposed to take place.

Clock 2 is driven positive, allowing charge to flow smoothly into the destination bucket. When clock 2 reaches its full positive value, all the charge is held on the destination bucket and no charge carriers are present in either the source bucket or the transfer switch channel region (if the transfer switch was turned on).

Clock 1 is driven positive, which has no effect if the transfer took place and simply allows the charge to return to its original state if the transfer did not take place.

Clock 3 is driven positive, allowing the charge on the transfer gate to return to the condition bucket.

Clock 4 is driven positive, opening the cut off switches that isolated the three buckets from outside influence during the reversible transfer operation.

This cycle of operations reversibly transfers charge from the source bucket to the destination bucket if the condition bucket holds a logic "0" (*e.g.*, no charge).

Energy Loss

There are losses during the RCT cycle of operations caused by current flowing through a resistive medium. The power lost will be, Power Loss = I^2 R where I is the current and R the resistance. Current is proportional to the frequency of operation of the circuit, so if the frequency is low the current will also be low. Therefore, power losses from resistive losses will fall off with the square of the frequency: Power Loss is proportional to f^2. Because the number of operations per second is also lower at a lower frequency, the actual energy loss per device operation is proportional to the frequency of operation: Energy Loss is proportional to f. By reducing the frequency of operation, the energy loss per operation can be reduced to whatever extent is desired. It should be remembered that "low frequency" operation is relative: CCD's with 0.9997 charge transfer efficiency operating at 1 gigahertz have been demonstrated.

More specifically, the electron drift velocity vd equals the electric field E times the mobility m:

$$vd = Em$$

If we let d be the distance traveled by an electron as it moves from one bucket to the next, and t be the time of a single logic operation, then it is approximately the case that:

$$d = vd\ t$$

The total energy dissipated by the movement of n electrons as they move from the source to the destination is simply the force times the distance times the number of electrons, or:

$$\text{Edissipated-neEd}$$

where Edissipated is the energy dissipated, n is the number of electrons in a packet, and e is the charge of an electron.

Putting these together yields:

$$\text{Edissipated} = ned^2/mt$$

If we assume that the distance traveled by the charge packet as it moves from one bucket to the next is.1 microns (10^{-5} centimeters), that one packet has 100 electrons, that the mobility is 1800 cm^2/Vs (the approximate mobility of diamond or silicon near room temperature), that the time allotted for one operation is 1 nanosecond, and the charge e on an electron is 1.6×10^{-19} coulombs, then the energy dissipated is approximately 10^{-21} joules.

This estimate is based on a classical model and involves some significant simplifications. Most significantly, the mean free path will of necessity be less than.1 micron in a device whose maximum dimension is.1 micron, and so the meaning of the electron mobility m is somewhat obscured.

A more detailed quantum analysis is essential as the size becomes smaller. A rough approximation to such an analysis can be based on the observation that the resistance of the channel of a sufficiently small FET (*e.g.*, one in which the channel width is perhaps a few nanometers and only a single transmission mode exists) is roughly half of h/e^2, where h is Planck's constant.

This is a fundamental unit of resistance of about 26K ohms. A packet of n electrons flowing in time t produces a current of ne/t. The power produced by such a current flow is $I^2 \times R$, while the energy dissipation per logic operation is $I^2 \times R \times t$, or $(ne/t)2 \times h/(2e^2) \times t$ which simplifies to:

$$\text{Edissipated} = hn^2/2t$$

For a switching time t of 1 nanosecond and a packet size n of 100 electrons, this is 3×10^{-21} joules for each packet that is passed through the channel of the FET. Under the specific conditions considered here this second approximation agrees at least in order of magnitude with the estimate based on a classical model. It is interesting to note that the two equations have significant differences.

The resistance of the channel in the classical case varies with the length of the channel. The length of the channel is less significant in the nonclassical case. The size of the charge packet, n, also influences energy dissipation in different ways in the two cases. In the classical case, the energy dissipation is a function of n rather than n^2.

In the classical case, two electrons drifting under the influence of the same electric field in separate regions of the channel will dissipate twice as much energy as a single electron. In effect, the resistance has been halved because the number of charge carriers has been doubled.

A similar effect would occur in a doped semiconductor if the doping density were doubled. If we assume a narrow channel that can only support a single transmission mode, however, then doubling the size of the charge packet will cause more electrons to move through the same channel in the same time period, with a resulting increase in interactions among the electrons.

The reader should note that no claim of universality or high accuracy is implied for either of the above equations. They are rough approximations that apply only under a limited range of conditions.

These conditions need not, in general, be true. In particular, it would seem attractive to design devices in which single electrons are always close to their ground state both during charge transport and during switching operations.

This would occur if the electron were always confined in a time varying potential well of relatively small dimensions. Devices with this property should be feasible when we are able to manufacture devices of sufficiently small size.

This should have a favourable influence on energy dissipation. The major factor ignored in this analysis is the energy dissipated by the clock signals. Many methods of providing periodic clocking pulses are possible. Three methods are discussed in the next section.

The mobility can be substantially increased by reducing the temperature. Mobilities greater than 1,000,000 have been demonstrated in GaAs-AlGaAs HFET's at around 10 K.

While the impact of this is unclear when the device size is much smaller than the mean free path (which occurs with such high mobilities), it does mean that larger devices could be made which would have very low energy dissipation.

The energy dissipation for this type of reversible electronic logic (as estimated here) is greater than the dissipation of Drexler's rod logic operated at a cycle time of 100 picoseconds (some 10 times faster than the 1 nanosecond considered here for the RCT). When RCT device operation using single electrons becomes feasible, then energy dissipation using RCT's would presumably be reduced.

Likharev's parametric quantron has an energy dissipation which (when adjusted for the temperature difference and the differing speeds of operation) is not as good as rod logic. The parametric quantron also requires low temperature.

Rod logic should work at room temperature. It is possible that mechanical molecular logic might have very favourable energy dissipation properties at low temperature.

While it is premature to draw conclusions it is not obvious that electronic devices must necessarily prove better than molecular mechanical devices in terms of energy dissipation at a given speed and temperature of operation.

Chapter 6

Nano Microbiology

Viral Envelope, Entry, Replication, Shedding and Latency

Viral Envelope

Many viruses (e.g. influenza and many animal viruses) have viral envelopes covering their protein capsids. The envelopes typically are derived from portions of the host cell membranes (phospholipids and proteins), but include some viral glycoproteins. Functionally, viral envelopes are used to help viruses enter host cells. Glycoproteins on the surface of the envelope serve to identify and bind to receptor sites on the host's membrane.

The viral envelope then fuses with the host's membrane, allowing the capsid and viral genome to enter and infect the host. Usually, the cell from which the virus itself buds will often die or be weakened, and shed more viral particles for an extended period. The lipid bilayer envelope of these viruses is relatively sensitive to desiccation, heat and detergents, therefore these viruses are easier to sterilize than non-enveloped viruses, have limited survival outside host environments, and typically must transfer directly from host to host.

Viral Entry

Viral entry is the earliest stage of infection in the viral life cycle, as the virus comes into contact with the host cell and introduces viral material into the cell. The major steps involved in viral entry are shown below. Despite the variation among viruses, the generalities are quite similar. However, the specifics are varied.

Reducing Cellular Proximity

A virus floating around an enclosed space with possible host cells faces a large hurdle, the thermodynamics of diffusion. Because neutrally charged objects do not naturally clump around each other, the virus must find a way to move even near a host cell. It does this by attachment — or adsorption — onto a susceptible cell; a cell which holds a receptor that the virus can bind to. The receptors on the viral envelope effectively become connected to complementary receptors on the cell membrane. This attachment causes the two membranes to remain in mutual proximity, favouring further interactions between surface proteins. This is also the first requisite that must be satisfied before a cell can become infected. Satisfaction of this requisite makes the cell susceptible. Viruses that exhibit this behaviour include many enveloped viruses such as HIV, Herpes simplex virus or influenza. This basic idea extends to viruses that do not contain an envelope. Well studied examples are the viruses that infect bacteria, known as bacteriophages or simply phages. Typical phages have long tails used to attach to receptors on the bacterial surface.

A virus must now enter the cell, which is covered by a phospholipid bilayer, a cell's natural barrier to the outside world. The process by which this barrier is breached depends upon the virus. Types of entry are:

1. *Attachment* or *Viral Adsorption*: Receptors on the viral envelope become connected to complementary receptors on the cell membrane.
2. *Membrane Fusion* or *Hemifusion State*: The cell membrane is punctured and made to further connect with the unfolding viral envelope.
3. *Entry Pore* formation: An opening is established for the stabilization of an opening for which viral particles can then enter.
4. *Viral Penetration*: The viral capsid or genome is injected into the host cell's cytoplasm.

Through the use of green fluorescent protein (GFP), virus entry and infection can be visualized in real-time. Once a virus enters a cell, replication is not immediate and indeed takes some time (seconds to hours).

Entry via Membrane Fusion

The most well-known example is through membrane fusion. In this case, viral receptors attach to the receptors on the surface of the cell and secondary receptors may be present to initiate the puncture of the cell membrane or fusion with the host cell, followed by the unfolding of the viral envelope. In essence, the virus's envelope blends with the cell membrane, releasing its contents into the cell. Obviously, this can only be done with viruses that contain an envelope. Examples include HIV, Herpes simplex virus or influenza.

Entry via Endocytosis

Another example is the triggered endocytosis of a cell, in which the virus tricks the cell into thinking that the virus knocking at the door is nothing more than nutrition or harmless goods. A cell naturally takes in resources from the environment and one of the ways it does so is by attaching goods onto surface receptors and engulfing them into the cell. Once inside the cell within the vacuole by which it was taken up, the virus must now break out of the vacuole to gain access to the cytoplasm. Examples include the poliovirus, Hepatitis C virus and Foot-and-mouth disease virus.

Entry via Genetic Injection

A third and more specific example, is by simply attaching to the surface of the cell via receptors on the cell, and injecting only its gene into the cell, leaving the rest of the virus on the surface. This is restricted to viruses in which only the gene is required for infection of a cell (most positive-sense, single-stranded RNA viruses) and further restricted to viruses that actually exhibit this behaviour. The best studied example includes the phages.

Aftermath

Once a virus is in a cell, it will activate formation of proteins (either by itself or using the host) to gain full control of the host cell, if it is able to. Control mechanisms include the suppression of intrinsic cell defenses, suppression of cell signaling and suppression of host cellular transcription and translation. Often, it is these cytotoxic effects that lead to the death and decline of a cell infected by a virus.

A cell is classified as susceptible to a virus if the virus is able to enter the cell. After the introduction of the viral particle, unpacking of the contents (viral proteins in the tegument and the viral genome via some form of nucleic acid) occurs as preparation of the next stage of viral infection: viral replication.

Viral Replication

Viral replication is the term used by virologists to describe the formation of biological viruses during the infection process in the target host cells. Viruses must first get into the cell before viral replication can occur. From the perspective of the virus, the purpose of viral replication is to allow production and survival of its kind. By generating abundant copies of its genome and packaging these copies into viruses, the virus is able to continue infecting new hosts. Replication between viruses is greatly varied and depends on the type of genes involved.

Baltimore Classification System

Viruses are classed into 7 types of genes, each of which has its own families of viruses, which in turn have differing replication strategies themselves. David Baltimore, a Nobel Prize-winning biologist, devised a system called the Baltimore Classification System to classify different viruses based on their unique replication strategy. There are seven different replication strategies based on this system (Baltimore Class I, II, III, IV, V, VI, VII). The seven classes of viruses are listed here briefly and in generalities.

Class 1: Double-Stranded DNA Viruses

This type of virus usually must enter the host nucleus before it is able to replicate. Some of these viruses require host cell polymerases to replicate their genome, while others, such as adenoviruses or herpes viruses, encode their own replication factors.

However, in either cases, replication of the viral genome is highly dependent on a cellular state permissive to DNA replication and, thus, on the cell cycle. The virus may induce the cell to forcefully undergo cell division, which may lead to transformation of the cell and, ultimately, cancer. An example of a family within this classification is the Adenoviridae. There is only one well-studied example in which a class 1 family of viruses does not replicate within the nucleus. This is the Poxvirus family, which comprises highly pathogenic viruses that infect vertebrates. One example is the smallpox virus.

Class 2: Single-Stranded DNA Viruses

Viruses that fall under this category include ones that are not as well-studied, but still do pertain highly to vertebrates. Two examples include the Circoviridae and Parvoviridae. They replicate within the nucleus, and form a double-stranded DNA intermediate during replication. A human Circovirus called TTV is included within this classification and is found in almost all humans, infecting them asymptomatically in nearly every major organ.

Class 3: Double-Stranded RNA Viruses

Like most viruses with RNA genomes, double-stranded RNA viruses do not rely on host polymerases for replication to the extent that viruses with DNA genomes do. Double-stranded RNA viruses are not as well-studied as other classes. This class includes two major families, the Reoviridae and Birnaviridae. Replication is monocistronic and includes individual, segmented genomes, meaning that each of the genes codes for only one protein, unlike other viruses, which exhibit more complex translation.

Classes 4 & 5: Single-Stranded RNA Viruses

These viruses consist of two types, however both share the fact that replication is primarily in the cytoplasm, and that replication is not as dependent on the cell cycle as that of DNA viruses. This class of viruses is also one of the most-studied types of viruses, alongside the double-stranded DNA viruses.

Class 4: Single-Stranded RNA viruses - Positive-Sense

The positive-sense RNA viruses and indeed all genes defined as positive-sense can be directly accessed by host ribosomes to immediately form proteins. These can be divided into two groups, both of which reproduce in the cytoplasm:

- Viruses with polycistronic mRNA where the genome RNA forms the mRNA and is translated into a polyprotein product that is subsequently cleaved to form the mature proteins. This means that the gene can utilize a few methods in which to produce proteins from the same strand of RNA, all in the sake of reducing the size of its gene.

- Viruses with complex transcription, for which subgenomic mRNAs, ribosomal frameshifting and proteolytic processing of polyproteins may be used. All of which are different mechanisms with which to produce proteins from the same strand of RNA.

Examples of this class include the families Coronaviridae, Flaviviridae, and Picornaviridae.

Class 5: Single-Stranded RNA Viruses - Negative-Sense

The negative-sense RNA viruses and indeed all genes defined as negative-sense cannot be directly accessed by host polymerases to immediately form proteins. Instead, they must be transcribed by viral polymerases into a "readable" form, which is the positive-sense reciprocal. These can also be divided into two groups:

- Viruses containing nonsegmented genomes for which the first step in replication is transcription from the negative-stranded genome by the viral RNA-dependent RNA polymerase to yield monocistronic mRNAs that code for the various viral proteins. A positive-sense genome copy that serves as template for production of the negative-strand genome is then produced. Replication is within the cytoplasm.
- Viruses with segmented genomes for which replication occurs in the nucleus and for which the viral RNA-dependent RNA polymerase produces monocistronic mRNAs from each genome segment. The largest difference between the two is the location of replication.

Class 6: Positive-Sense Single-Stranded RNA Viruses that Replicate through a DNA Intermediate

A well-studied family of this class of viruses include the retroviruses. One defining feature is the use of reverse transcriptase to convert the positive-sense RNA into DNA. Instead of using the RNA for templates of proteins, they use DNA to create the templates, which is spliced into the host genome using integrase. Replication can then commence with the help of the host cell's polymerases. A well-studied example includes HIV.

Class 7: Double-Stranded DNA Viruses that Replicate through a Single-Stranded RNA Intermediate

This small group of viruses, exemplified by the Hepatitis B virus, have a double-stranded, gapped genome that is subsequently filled in to form a covalently closed circle (ccc DNA) that serves as a template for production of viral mRNAs and a subgenomic RNA. The pregenome RNA serves as template for the viral reverse transcriptase and for production of the DNA genome.

Viral Shedding

Viral shedding refers to the successful reproduction, expulsion, and host-cell infection caused by virus progeny. Once replication has been completed and the host cell is exhausted of all resources in making viral progeny, the viruses may begin to leave the cell by several methods. The term is used to refer to shedding from a single cell, shedding from one part of the body into another part of the body, and shedding from bodies into the environment where the viruses may infect other bodies.

Via Budding

"Budding" through the cell envelope, in effect using the cell's membrane for the virus itself is most effective for viruses that need an envelope in the first place. These include enveloped viruses such as HSV, SARS or smallpox. Prior to budding, the virus may put its own receptor onto the surface of the cell in preparation for the virus to bud through, forming an envelope with the viral receptors already on it. This process will slowly use up the cell membrane and eventually lead to the demise of the cell. This is also how antiviral responses are able to detect virus infected cells.

Via Apoptosis

By forcing the cell to undergo apoptosis or cell suicide, release of progeny into the extracellular space is possible. However, apoptosis does not necessarily result in the cell simply popping open and spilling its contents into the extracellular space. Rather, apoptosis is usually controlled and results in the cell's genome being chopped up, before apoptotic bodies of dead cell material clump off the cell to be absorbed by macrophages. This is a good way for a virus to get into macrophages either to infect them or simply travel to other tissues in the body. Although this process is primarily used by non-enveloped viruses, enveloped viruses may also use this. HIV is an example of an enveloped virus that exhibits this process for the infection of macrophages.

Via Exocytosis

It may not be in a virus's best interest to kill the cell in order to escape it for further infection of other host cells. In such a case, exocytotic release of viral progeny is used to release viral particles. Viral progeny are synthesized within the cell and the host cell's transport system is used to enclose vacuoles of virus progeny for release into the extracellular space. This is used primarily by non-enveloped viruses, although enveloped viruses display this too. An example is the use of recycling viral particle receptors in the enveloped varicella-zoster virus.

Virus Latency

Virus latency (or viral latency) is the ability of a pathogenic virus to lie dormant (latent) within a cell, denoted as the lysogenic part of the viral life cycle. A latent viral infection is a type of persistent viral infection which is distinguished from a chronic viral infection. A latent infection is a phase in certain viruses' life cycles in which after initial infection, virus production ceases. However, the virus genome is not fully eradicated. The result of this is that the virus can reactivate and begin producing large amounts of viral progeny without the host being infected by new outside virus, denoted as the lytic part of the viral life cycle, and stays within the host indefinitely. Virus latency is not to be confused with clinical latency during the incubation period when a virus is not dormant.

Mechanisms

Episomal Latency: Episomal latency refers to the use of genetic episomes during latency. In this type, viral genes are floating in the cytoplasm or nucleus as distinct objects, both as linear or lariat structures. Episomal latency is more vulnerable to ribozymes or host foreign gene degradation than provirus latency. One example is Herpes Virus family, *Herpesviridae*, all of which establish latent infection. Herpes virus include Chicken-pox virus and Herpes simplex viruses (HSV-1, HSV-2), all of which establish episomal latency in neurons and leave linear genetic material floating in the cytoplasm. The *Gammaherpesvirinae* subfamily is associated with episomal latency established in cells of the immune system, such as B-cells in the case of Epstein-Barr Virus.

Advantages of episomal latency include the fact that the virus may not need to enter the nucleus, and hence may avoid ND10 domains from activating interferon via that pathway. Disadvantages include more exposure to cellular defenses, leading to possible degradation of viral gene via cellular enzymes.

Proviral Latency

Proviral latency begins when the virus genome integrates into the host genome, effectively become a provirus. This requires that the viral gene get into the nucleus and insert itself into the host genome, the family of which exemplifies this behaviour being the Retroviruses. For example, the retrovirus HIV enters the nucleus and inserts its gene between Long terminal repeats using integrase and remains within the hosts own gene. Advantages include automatic host cell division results in replication of the viruses gene, and the fact that it is near impossible to remove an integrated provirus from an infected cell without killing the cell. Disadvantages include the need to enter the nucleus (and the need for packaging proteins that will allow for that) and increased difficulty in maintaining the latency.

Maintaining Latency

Both proviral and episomal latency may require maintenance for continued infection and fidelity of viral genes. Latency is generally maintained by viral genes expressed primarily during latency. Expression of these *latency-associated* genes may function to keep the viral genome from being digested by cellular ribozymes or being found out by the immune system. Certain viral gene products (RNA transcripts such as non-coding RNAs and proteins) may also inhibit apoptosis or induce cell growth and division to allow more copies of the infected cell to be produced. An example of such a gene product is the *Latency Associated Transcripts (LAT)* in Herpes simplex virus, which interfere with apoptosis by downregulating a number of host factors, including Major Histocompatibility Complex (MHC) and inhibiting the apoptotic pathway.

A certain type of latency could be ascribed to the endogenous retroviruses. These viruses have incorporated into the human genome in the distant past, and are now passed through reproduction. Generally these types of viruses have become highly evolved, and have lost the expression of many gene products. Some of the proteins expressed by these viruses have co-evolved with host cells to play important roles in normal processes.

Ramifications

While viral latency exhibits no active viral shedding nor causes any pathologies or symptoms, the virus is still able to reactivate via external activators (i.e. sunlight, stress) to cause an acute infection. In the case of Herpes simplex virus, which generally infects an individual for life, a serotype of the virus reactivates occasionally to cause cold sores. Although the sores are quickly resolved by the immune system, they may be a minor annoyance from time to time.

In the case of varicella zoster virus, after an initial acute infection (chickenpox) the virus lies dormant until reactivated as herpes zoster. More serious ramifications of a latent infection could be the possibility of transforming the cell, and forcing the cell into uncontrolled cell division. This is a result of the random insertion of the viral genome into the hosts own gene and suppression of host cellular growth factors for the benefit of the virus.

A famous event of this actually happening with gene therapy through the use of retroviral vectors is the Necker Hospital in Paris, where 8 young boys received treatment for a genetic disorder, after which 4 developed leukemia. This is also seen with infections of the human papilloma virus in which persistent infection may lead to cervical cancer as a result of cellular transformation. In the field of HIV research, proviral latency in specific long-lived cell types is the basis for the concept of one or more viral reservoirs, referring to locations (cell types or tissues) characterized by persistence of latent virus. Specifically, the presence of replication-competent HIV in resting CD4-positive T cells, allows this virus to persist for years without evolving despite prolonged exposure to antiretroviral drugs. This latent reservoir of HIV may explain the inability of antiretroviral treatment to cure HIV infection.

Sporangium

A sporangium (pl., sporangia) (modern Latin, from Greek spora 'spore' + angeion 'vessel.') is an enclosure in which spores are formed. It can be composed of a single cell or can be multicellular. All plants, fungi, and many other lineages form sporangia at some point in their life cycle. Sporangia, also known as sporanges (sing., sporange), can produce spores by mitosis, but in nearly all land plants and many fungi, sporangia are the site of meiosis and produce genetically unique haploid spores.

Land Plants

In mosses, liverworts and hornworts, an unbranched sporophyte produces a single sporangium, which may be quite complex morphologically. Most non-vascular plants, as well as many lycophytes and most ferns, are homosporous (only one kind of spore is produced). Some bryophytes, most lycophytes, and some ferns are heterosporous (two kinds of spores are produced).

These plants produce microspores and megaspores, which give rise to gametophytes that are functionally male or female, respectively. In some cases, both kinds of spores are produced in the same sporangium, and may even develop together as part of a spore tetrad.

However, in most heterosporous plants there are two kinds of sporangia, termed microsporangia and megasporangia. A few ferns (Salviniaceae and Marsileaceae) and most lycophytes (the genera *Selaginella* and *Isoetes* and the extinct lepidodendrids) are heterosporous with two kinds of sporangia, as are all the seed plants.

Sporangia can be terminal (on the tips) or lateral (placed along the side) of stems or associated with leaves. In ferns, sporangia are typically found on the abaxial surface (underside) of the leaf and are densely aggregated into clusters called sori. Sori may be covered by a structure called an indusium. Some ferns have their sporangia scattered along reduced leaf segments or along (or just in from) the margin of the leaf. Lycophytes, in contrast, bear their sporangia on adaxial surface (the upper side) of leaves or laterally on stems. Leaves that bear sporangia are called sporophylls.

If the plant is heterosporous, the sporangia-bearing leaves are distinguished as either microsporophylls or megasporophylls. In seed plants, sporangia are typically located within strobili or flowers. Cycads place their microsporangia on micosporophylls which are aggregated into strobili. Megasporangia are similarly found on megasporophylls, which are aggregated into strobili on separate plants (all cycads are dioecious).

Conifers typically bear their micosporangia on microsporophylls aggregated into papery pollen strobili while megasporangia are inside ovules which are located on modified stem axes forming compound ovuliferous cone scales. Flowering plants contain microsporangia in the anthers of stamens (typically four microsporangia per anther) and megasporangia inside ovules inside ovaries.

In all seed plants, spores are produced by meiosis and develop into gametophytes while still inside the sporangium. The microspores become microgametophytes (pollen). The megaspores become megagametophytes (embryo sacs).

Eusporangia and Leptosporangia

Categorized based on developmental sequence, *eusporangia* and *leptosporangia* are differentiated in the vascular plants.

- In a leptosporangium, found only in ferns, development involves a single initial cell that becomes the stalk, wall, and spores within the sporangium. There are around 64 spores in a leptosporangium.
- In a eusporangium, characteristic of all other vascular plants and some primitive ferns, the initials are in a layer (i.e., more than one). A eusporangium is larger (hence contain more spores), and its wall is multi-layered. Although the wall may be stretched and damaged, resulting in only one cell-layer remaining.

Synangium

A cluster of sporangia that have become fused in development is called a synangium. This structure is most prominent in *Psilotum* and Marattiaceae such as *Christensenia*, *Danaea* and *Marattia*.

Internal Structures

A columella (pl. columellae) is a sterile (non-reproductive) structure that extends into and supports the sporangium. In fungi, the columella, which may be branched or unbranched, may be of fungal or host origin.

Secotium species have a simple, unbranched columella, while in *Gymnoglossum* species, the columella is branched. In some *Geastrum* species, the columella appears as an extension of the stalk into the spore mass (gleba).

Conidium

Conidia, sometimes termed conidiospores, are asexual, non-motile spores of a fungus and are named after the greek word for dust, konia. They are also called mitospores due to the way they are generated through the cellular process of mitosis.

The two new haploid cells are genetically identical to the haploid parent, and can develop into new organisms if conditions are favourable, and serve in biological dispersal.

Asexual reproduction in Ascomycetes (the Phylum Ascomycota) is by the formation of conidia, which are borne on specialized stalks called conidiophores. The morphology of these specialized conidiophores is often distinctive of a specific species and can therefore be used in identification of the species.

The terms "microconidia" and "macroconidia" are sometimes used.

Conidiogenesis

There are two main types of conidium development:

- *Blastic* conidiogenesis, where the spore is already evident before it separates from the conidiogenic hypha which is giving rise to it, and
- *Thallic* conidiogenesis, where first a cross-wall appears and then the thus created cell develops into a spore.

Conidia Germination

A conidium may form germ tubes (germination tubes) and/or conidial anastomosis tubes (CATs) in specific conditions. These two are some of the specialized hyphae that are formed by fungal conidia. The germ tubes will grow to form the hyphae and fungal mycelia. The conidial anastomosis tubes are morphologically and physiologically distinct from germ tubes.

After conidia are induced to form conidial anastomosis tubes, they grow homing toward each other, and they fuse. Once fusion happens, the nuclei can pass through fused CATs. These are events of fungal vegetative growth and not sexual reproduction.

Fusion between these cells seems to be important for some fungi during early stages of colony establishment. The production of these cells has been suggested to occur in 73 different species of fungi.

Structures for Release of Conidia

Conidiogenesis is an important mechanism of spread of plant pathogens. In some cases, specialized macroscopic fruiting structures perhaps 1mm or so in diameter containing masses of conidia are formed under the skin of the host plant and then erupt through the surface and allow the spores to be distributed by wind and rain. One of these structures is called a conidioma (plural: conidiomata).

Two important types of conidiomata, distinguished by their form, are:

- pycnidia, (singular: pycnidium) which are flask-shaped, and
- acervuli (singular: acervulus), which have a simpler cushion-like form.

Pycnidial conidiomata or *pycnidia* form in the fungal tissue itself, and are shaped like a bulging vase. The spores are released through a small opening at the apex, the ostiole.

Acervular conidiomata, or *acervuli*, are cushion-like structures that form within the tissues of a host organism:

- *subcuticular*, lying under the outer layer of the plant (the cuticle),

- *intraepidermal*, inside the outer cell layer (the epidermis),
- *subepidermal*, under the epidermis, or deeper inside the host.

Mostly they develop a flat layer of relatively short conidiophores which then produce masses of spores. The increasing pressure leads to the splitting of the epidermis and cuticle and allows release of the conidia from the tissue.

Health Issues

Conidia are always present in the air, but levels fluctuate from day to day and with the seasons. An average person inhales 40 conidia per hour.

Conidia are often the method by which some normally harmless but heat-tolerating (thermotolerant), common fungi establish infection in certain types of severely immunocompromised patients (usually acute leukemia patients on induction chemotherapy, AIDS patients with superimposed B-cell lymphoma, bone marrow transplantation patients, or major organ transplant patients suffering from graft versus host disease).

Their immune system is not strong enough to fight off the fungus, and it may, for example, colonise the lung, resulting in a pulmonary infection.

Nitrogen Fixation

Nitrogen fixation is the natural process, either biological or abiotic, by which nitrogen (N_2) in the atmosphere is converted into ammonia (NH_3). This process is essential for life because fixed nitrogen is required to biosynthesize the basic building blocks of life, e.g., nucleotides for DNA and RNA and amino acids for proteins. Nitrogen fixation also refers to other biological conversions of nitrogen, such as its conversion to nitrogen dioxide.

Microorganisms that fix nitrogen are bacteria called diazotrophs. Some higher plants, and some animals (termites), have formed associations (symbioses) with diazotrophs. Nitrogen fixation also occurs as a result of non-biological processes. These include lightning, industrially through the Haber-Bosch

Process, and combustion. Biological nitrogen fixation was discovered by the Dutch microbiologist Martinus Beijerinck.

Biological Nitrogen Fixation

Biological nitrogen fixation (BNF) occurs when atmospheric nitrogen is converted to ammonia by an enzyme called nitrogenase. The reaction for BNF is:

$$N_2 + 8\ H^+ + 8\ e^- \longrightarrow 2\ NH_3 + H_2$$

The process is coupled to the hydrolysis of 16 equivalents of ATP and is accompanied by the co-formation of one molecule of H_2. In free-living diazotrophs, the nitrogenase-generated ammonium is assimilated into glutamate through the glutamine synthetase/glutamate synthase pathway.

Enzymes responsible for nitrogenase action are very susceptible to destruction by oxygen. (In fact, many bacteria cease production of the enzyme in the presence of oxygen). Many nitrogen-fixing organisms exist only in anaerobic conditions, respiring to draw down oxygen levels, or binding the oxygen with a protein such as Leghemoglobin.

Microorganisms that Fix Nitrogen (Diazotrophs)

- Cyanobacteria
- *Azotobacteraceae*
- *Rhizobia*
- *Frankia*

Nitrogen Fixation by Rhizobia

Rhizobia are soil, Gram-negative bacteria with the unique ability to establish a N_2-fixing symbiosis on legume roots and on the stems of some aquatic legumes. During this interaction bacteroids, as rhizobia are called in the symbiotic state, are contained in intracellular compartments within a specialized organ, the nodule, where they fix N_2.

Nitrogen Fixation by Cyanobacteria

Cyanobacteria inhabit nearly all illuminated environments on Earth and play key roles in the carbon and nitrogen cycle of the biosphere. In general, cyanobacteria are able to utilize a variety of inorganic and organic sources of combined nitrogen, like nitrate, nitrite, ammonium, urea, or some amino acids.

Several cyanobacterial strains are also capable of diazotrophic growth. Genome sequencing has provided a large amount of information on the genetic basis of nitrogen metabolism and its control in different cyanobacteria. Comparative genomics, together with functional studies, has led to a significant advance in this field over the past years. 2-Oxoglutarate has turned out to be the central signalling molecule reflecting the carbon/nitrogen balance of cyanobacteria.

Central players of nitrogen control are the global transcriptional factor NtcA, which controls the expression of many genes involved in nitrogen metabolism, as well as the P_{II} signalling protein, which fine-tunes cellular activities in response to changing C/N conditions. These two proteins are sensors of the cellular 2-oxoglutarate level and have been conserved in all cyanobacteria. In contrast, the adaptation to nitrogen starvation involves heterogeneous responses in different strains. Nitrogen fixation by cyanobacteria in coral reefs can fix twice the amount of nitrogen than on land–around 1.8 kg of nitrogen is fixed per hectare per day.

Symbiotic Nitrogen Fixing Plants

Legume Family: Plants that contribute to nitrogen fixation include the legume family – Fabaceae – with taxa such as clover, soybeans, alfalfa, lupines, peanuts, and rooibos.

They contain symbiotic bacteria called *Rhizobia* within nodules in their root systems, producing nitrogen compounds that help the plant to grow and compete with other plants. When the plant dies, the fixed nitrogen is released, making it available to other plants and this helps to fertilize the soil The great majority of legumes have this association, but a few genera (e.g., *Styphnolobium*) do not.

In many traditional and organic farming practices, fields are rotated through various types of crops, which usually includes one consisting mainly or entirely of clover or buckwheat (family *Polygonaceae*), which were often referred to as "green manure."

Non-Leguminous

Although by far the majority of nitrogen-fixing plants are in the legume family Fabaceae, there are a few non-leguminous plants, such as alder and bayberry, that can also fix nitrogen, thanks to a symbiotic associatioin with Frankia bacteria.

These plants, referred to as "actinorhizal plants", consist of 24 genera of woody shrubs or trees distributed among 8 plant families. The ability to fix nitrogen is far from universally present in these families. For instance, of 122 genera in the Rosaceae, only 4 genera are capable of fixing nitrogen. All these families belong to the orders Cucurbitales, Fagales, and Rosales, which together with the Fabales form a clade of eurosids.

Chemical Nitrogen Fixation

Haber Process: Nitrogen can also be artificially fixed as ammonia for use in fertilizers, explosives, or in other products. The most common method is the Haber process. Artificial fertilizer production is now the largest source of human-produced fixed nitrogen in the Earth's ecosystem. The Haber process requires high pressures (around 200 atm) and high temperatures (at least 400 °C), routine conditions for industrial catalysis. This highly efficient process uses natural gas as a hydrogen source and air as a nitrogen source.

Dinitrogen Complexes

Much research has been conducted on the discovery of catalysts for nitrogen fixation, often with the goal of reducing the energy required for this conversion. However, such research has thus far failed to even approach the efficiency and ease of the Haber process. Many compounds react with atmospheric nitrogen under ambient conditions. For example, lithium metal converts to lithium nitride under an atmosphere of nitrogen. Treatment of the resulting nitride gives ammonia.

The first dinitrogen complex was reported in 1965 based on ammonia coordinated to ruthenium ($[Ru(NH_3)_5(N_2)]^{2+}$). Research in chemical fixation from then on focused on transition metal complexes. Since then, a large number of transition metal compounds that contain dinitrogen as a ligand have been discovered. The dinitrogen ligand can either be bound to a single metal or bridge two (or more) metals. The coordination chemistry of dinitrogen is complex and currently under intense investigation. This research may lead to new ways of using dinitrogen in synthesis and on an industrial scale.

Ambient Nitrogen Reduction

Catalytic chemical nitrogen fixation at temperatures considerably lower than the Haber process is an ongoing scientific endeavor. Nitrogen was successfully converted to ammonia and hydrazine by Alexander E. Shilov in 1970 The first example of homolytic cleavage of dinitrogen under mild conditions was published in 1995. Two equivalents of a molybdenum complex reacted with one equivalent of dinitrogen, creating a triple bonded MoN complex. Since then, this triple bounded complex has been used to make nitriles.

In 2011 Arashiba et al. reported another system with a catalyst again based on molybdenum but with a diphosphorus pincer ligand

Acetic Acid Bacteria

Acetic acid bacteria (AAB) are bacteria that derive their energy from the oxidation of ethanol to acetic acid during fermentation. They are Gram-negative, aerobic, rod-shaped bacteria. Not to be confused with the genus *Acetobacterium* which are anaerobic homoacetogenic facultative autotrophs and can reduce carbon dioxide to produce acetic acid.

For example: *Acetobacterium woodii.* The acetic acid bacteria are usually airborne and are ubiquitous in nature. They are actively present in environment where ethanol is being formed as a result of fermentation of sugars. They can be isolated from the nectar of flowers and from damaged fruit.

Other good sources are fresh apple cider and unpasteurized beer which has not been filter sterilized. In these liquids, they grow as a surface film due to their aerobic nature and active motility. Vinegar is produced when acetic acid bacteria act on alcoholic beverages such as wine. Fruit flies or Vinegar eels are considered as a common vector in propagating acetic acid bacteria in nature. Some genera, such as *Acetobacter*, can eventually oxidize acetic acid to carbon dioxide and water using Krebs cycle enzymes. Other genera, such as *Gluconobacter*, do not further oxidize acetic acid, as they do not have a full set of Krebs cycle enzymes.

Some acetic acid bacteria, notably *Acetobacter xylinum*, are known to synthesize cellulose, something normally only done by plants. As these bacteria produce acid, they are usually acid tolerant, growing well below pH 5.0, although the pH optimum for growth is 5.4-6.3.

Acetogen

An acetogen is a microorganism that generates acetate as a product of anaerobic respiration. This process is known as acetogenesis and is different from acetate fermentation, although both occur in the absence of oxygen and produce acetate. These organisms are also referred to as *acetogenic bacteria,* because all known acetogens are bacteria.

Acetogens are found in a variety of habitats, generally those that are anaerobic (lack oxygen). Acetogens can use a variety of compounds as sources of energy and carbon; the best studied form of acetogenic metabolism involves the use of carbon dioxide as a carbon source and hydrogen as an energy source.

Acid-Fast

Acid-fastness is a physical property of some bacteria referring to their resistance to decolorization by acids during staining procedures. Acid-fast organisms are difficult to characterize using standard microbiological techniques (e.g. Gram stain - if you gram stained an AFB the result would be an abnormal gram positive organism, which would indicate further testing), though they can be stained using concentrated dyes, particularly when the staining process is combined with heat. Once stained, these organisms resist the dilute acid and/or ethanol-based de-colorization procedures common in many staining protocols—hence the name *acid-fast.*

The high mycolic acid content of certain bacterial cell walls, like those of *Mycobacteria,* is responsible for the staining pattern of poor absorption followed by high retention. The most common staining technique used to identify acid-fast bacteria is the Ziehl-Neelsen stain, in which the acid fast bacilli are stained bright red and stand out clearly against a blue background.

Another method is the Kinyoun method, in which the bacteria are stained bright red and stand out clearly against a green background. Acid-fast bacteria can also be visualized by fluorescence microscopy using specific fluorescent dyes (auramine-rhodamine stain, for example). Some bacteria may also be partially acid-fast.

Notable Acid Fast Structures

Very few structures are acid fast; this makes staining for acid-fastness particularly useful in diagnosis.

- All Mycobacteria - *M. tuberculosis*, *M. leprae*, *M. smegmatis* and atypical *Mycobacterium*
- *Nocardia*
- Head of sperm
- Bacterial spores, see Endospore
- Some coccidian parasites, such as *Cryptosporidium parvum*, *Isospora belli* and *Cyclospora cayetanensis* cysts in faecal matter.

Acidobacteria

Acidobacteria are a newly devised phylum of Bacteria, whose members are physiologically diverse and ubiquitous, especially in soils, but are under-represented in culture. Members of this phylum are physiologically diverse, some being acidophilic, and were first recognized as a novel division in 1997.

The first species, *Acidobacterium capsulatum*, of this phylum was discovered in 1991. Other notable species are *Holophaga foetida*, *Geothrix fermentans*, *Acanthopleuribacter pedis* and *Bryobacter aggregatus*. Since they have only recently been discovered and the large majority have not been cultured, the ecology and metabolism of these bacteria is not well understood. However, these bacteria may be an important contributor to ecosystems, since they are particularly abundant within soils.

Subdivision

Current classification subdivides the phylum as follows:

- Class Acidobacteria
  - Order Acidobacteriales
    - Family Acidobacteriaceae (*Acidobacterium* - *Edaphobacter*- *Granulicella* - *Terriglobus*)

- Class Holophagae
  - Order Acanthopleuribacterales
    - Family Acanthopleuribacteraceae (*Acanthopleuribacter*)
  - Order Holophagales
    - Family Holophagaceae (*Geothrix* - *Holophaga*)
  - Unclassified Acidobacteria *Bryobacter*

Acidogenesis

Acidogenesis represents the second stage in the four stages of anaerobic digestion:

- *Hydrolysis:* A chemical reaction where particulates are solubilized and large polymers converted into simpler monomers;
- *Acidogenesis:* A biological reaction where simple monomers are converted into volatile fatty acids;
- *Acetogenesis:* A biological reaction where volatile fatty acids are converted into acetic acid, carbon dioxide, and hydrogen; and
- *Methanogenesis:* A biological reaction where acetates are converted into methane and carbon dioxide, while hydrogen is consumed.

Anaerobic digestion is a complex biochemical process of biologically mediated reactions by a consortium of microorganisms to convert organic compounds into methane and carbon dioxide. It is a stabilization process, reducing odor, pathogens, and mass reduction. Hydrolytic bacteria form a variety of reduced end-products from the fermentation of a given substrate.

One fundamental question that arises concerns the metabolic features that control carbon and electron flow to a given reduced end-product during pure culture and mixed methanogenic cultures of hydrolytic bacteria.

Thermoanaerobium brockii is a representative thermophilic, hydrolytic bacterium, which ferments glucose, via the Embden-Meyerhof Parnas Pathway. *T. brockii* is an atypical hetero-lactic acid bacterium because it forms molecular hydrogen (H_2), in addition to lactic acid and ethanol. The reduced end-products of glucose fermentation are enzymatically formed from pyruvate, via the following mechanisms: lactate by all-phosphate (F6P) activated lactate dehydrogenase; H2 by pyruvate ferredoxin oxidoreductase and hydrogenase; and ethanol via NADH- and NADPH-linked alcohol dehydrogenase.

By its side, the acidogenic activity was found in the early 20th century, but it was not until mid-'60s that the engineering of phases separation was assumed in order to improve the stability and waste digesters treatment.

In this phase, complex molecules (carbohydrates, lipids, and proteins) are depolymerized into soluble compounds by hydrolytic enzymes (cellulases, hemicellulases, amylases, lipases and proteases). The hydrolyzed compounds are fermented into volatile fatty acids (acetate, propionate, butyrate, and lactate), neutral compounds (ethanol, methanol), ammonia, hydrogen and carbon dioxide.

Acetogenesis is one of the main reactions of this stage, in this, the intermediary metabolites produced are metabolized to acetate, hydrogen and carbonic gas by the three main groups of bacteria:

- homoacetogens;
- syntrophes; and
- sulphoreductors.

For the acetic acid production are considered three kind of bacteria:

- *Clostridium aceticum*;
- *Acetobacter woodii*; and
- *Clostridium termoautotrophicum*.

Winter y Wolfe, in 1979, demonstrated that *A. wodii* in syntrophic association with *Methanosarcina* produce methane and carbon dioxide from fructose, instead of three molecules of acetate. *C. thermoaceticum* and *C. formiaceticum* are able to reduce the carbonic gas to acetate, but they do not have hydrogenases which inhabilite the hydrogen use, so they can produce three molecules of acetate from fructose. Acetic acid is equally a co-metabolite of the organic substrates fermentation (sugars, glycerol, lactic acid, etc.) by diverse groups of microorganisms which produce different acids: *propionic bacteria (propionate + acetate);

- *Clostridium* (butyrate + acetate);
- Enterobacteria (acetate + lactate); and
- Hetero-fermentative bacteria (acetate, propionate, butyrate, valerate, etc.).

Acinetobacter Baumannii

Acinetobacter baumannii is a species of pathogenic bacteria, referred to as an aerobic gram-negative bacterium, that is resistant to most antibiotics. As a result of its resistance to drug treatment, some estimates state the disease is killing tens of thousands of U.S. hospital patients each year, and specialists say "they could emerge as a bigger threat." The illness can cause severe pneumonia and infections of the urinary tract, bloodstream and other parts of the body.

A. baumannii forms opportunistic infections. There have been many reports of *A. baumannii* infections among American soldiers wounded in Iraq, earning it the nickname "Iraqibacter". Multidrug-resistant *Acinetobacter baumannii* is abbreviated as MDRAB. Multidrug-resistant Acinetobacter is not a new phenomenon; it has always been inherently resistant to multiple antibiotics.

Biology

Acinetobacter baumannii is the most relevant human pathogen within the *Acinetobacter* genus. Most *A. baumannii* isolates are multiresistant, containing in their genome small, isolated islands of alien (meaning transmitted genetically from other organisms) DNA and other cytological and genetic material; this has led to more virulence. *Acinetobacter* have no flagellum; the name is Greek for "motionless".

During growth, they are rod-shaped, and up to 2.5 microns; but, during the stationary phase, they become shorter and rounder, almost like little cocci, as they shrink to roughly half the size. They can be found in pairs or groups. One of the tests that distinguish them from other pathogens is that they are oxidase-negative.

Transmission and Prevalence

Acinetobacter enters into the body through open wounds, catheters, and breathing tubes. It usually infects those with compromised immune systems, such as the wounded, the elderly, children, or those with immune diseases.

Colonization poses no threat to people not already ill, but colonized healthcare workers and hospital visitors can carry the bacteria into neighboring wards and other medical facilities. The number of nosocomial infections (hospital-acquired infections) caused by *A. baumannii* has increased in recent years, as have most other nosocomial pathogens (MRSA, VRSA, VRE, etc.) One of the reasons is that the bacteria can live up to 5 months on undisturbed surfaces, depending on humidity levels.

The first military outbreaks of severe *A. baumannii* infections occurred in April, 2003 in American soldiers returning from Iraq. Early reports attributed the infections to the Iraqi soil. Later testing demonstrated widespread contamination of field hospitals, via transportation of personnel and equipment from previously contaminated European hospitals, as the most plausible vector.

Virulence and Pathogenicity

This opportunistic pathogen causes a wide variety of serious infections in humans, mostly in compromised patients. Recently, *A. baumannii* has emerged as an important pathogen among wounded soldiers, threatening civilian and military patients. A baumannii has also been implicated in severe life threatening infections such as necrotizing fasciitis.

Nosocomial *A. baumannii* bacteremia may cause severe clinical disease that is associated with an elevated mortality rate. This opportunistic pathogen expresses myriad factors that could play a role in human pathogenesis.

Among these factors are the attachment to and persistence on solid surfaces, the acquisition of essential nutrients such as iron, the adhesion to epithelial cells and their subsequent killing by apoptosis, and the production and/or secretion of enzymes and toxic products that damage host tissues. However, very little is known about the molecular nature of most of these processes and factors, and almost nothing has been shown with regard to their role in bacterial virulence and the pathogenesis of serious infectious diseases.

However, some of these gaps can now be filled by testing appropriate isogenic derivatives in relevant animal models that mimic the infections in humans, in particular the outcome of deadly pneumonia. Such an approach should provide new and relevant information on the virulence traits of this normally underestimated bacterial human pathogen.

Control and Treatment of Infections

Multidrug-resistant *A. baumannii* is a common problem in many hospitals in the US and Europe. First-line treatment is with a carbapenem antibiotic such as imipenem, but carbapenem resistance is increasingly common. Other treatment options include polymyxins, tigecycline, and aminoglycosides.

The institution of strict infection-control measures, such as monitored hand washing, can lower hospital infection rates. MDRAB infections are difficult and costly to treat. A study at a public teaching hospital found that the mean total hospital cost of patients having acquired MDRAB was $98,575 higher than that of control patients having had identical burn severity of illness indices. There are some specific treatments in developing stage in order to overcome the resistance of this bacterium, using a specific bacteriophage against this type of bacteria.

Actinobacillus

Actinobacillus is a genus of gram-negative, immotile and nonspore-forming, oval to rod-shaped bacteria occurring as parasites or pathogens in mammals, birds, and reptiles. It is a member of the *Pasteurellaceae* family. The bacteria are facultatively aerobic, capable of fermenting carbohydrates (without production of gas) and of reducing nitrates. The genomic DNA contains between 40 and 47 mol % guanine plus cytosine.

Actinobacillus (Pasteurella) ureae and *A. hominis* occur in the respiratory tract of healthy humans and may be involved in the pathogenesis of sinusitis, bronchopneumonia, and meningitis. *Actinobacillus actinomycetemcomitans* occurs in the human oral microflora, and together with anaerobic or capnophilic organisms (HACEK group organisms) may cause endocarditis. Actinobacilli are susceptible to most antibiotics of the beta-lactam family, aminoglycosides, tetracyclines, chloramphenicol, and many other antibacterial chemotherapeutics.

A analysis of *Actinobacillus actinomycetemcomitans* indicated that it was monophyletic with *Haemophilus aphrophilus* and *Haemophilus segnis*, and it was proposed that they be reclassified as a new genus, Aggregatibacter (from the Latin, "aggregare", meaning "to come together").

Actinobacteria

Actinobacteria are a group of Gram-positive bacteria with high G+C ratio. They can be terrestrial or aquatic. Actinobacteria is one of the dominant phyla of the bacteria.

Use of the ferric uptake regulator (fur) has been suggested for classification. Analysis of glutamine synthetase has also been suggested. A Siberian example is said to be the oldest living organism on Earth, at around half a million years old.

Characteristics

They include some of the most common soil life, freshwater life, and marine life, playing an important role in decomposition of organic materials, such as cellulose and chitin, and thereby playing a vital part in organic matter turnover and carbon cycle.

This replenishes the supply of nutrients in the soil and is an important part of humus formation. Other Actinobacteria inhabit plants and animals, including a few pathogens, such as *Mycobacterium*, *Corynebacterium*, *Nocardia*, *Rhodococcus* and a few species of *Streptomyces*.

Actinobacteria are well known as secondary metabolite producers and hence of high pharmacological and commercial interest. In 1940 Selman Waksman discovered that the soil bacteria he was studying made actinomycin, a discovery for which he received a Nobel Prize. Since then, hundreds of naturally occurring antibiotics have been discovered in these terrestrial microorganisms, especially from the genus *Streptomyces*.

Some Actinobacteria form branching filaments, which somewhat resemble the mycelia of the unrelated fungi, among which they were originally classified under the older name Actinomycetes.

Most members are aerobic, but a few, such as *Actinomyces israelii*, can grow under anaerobic conditions. Unlike the Firmicutes, the other main group of Gram-positive bacteria, they have DNA with a high GC-content, and some Actinomycetes species produce external spores.

Some types of Actinobacteria are responsible for the peculiar odor emanating from the soil after rain (Petrichor), mainly in warmer climates. The chemical that produces this odour is known as Geosmin.

Genera

Most Actinobacteria of medical or economic significance are in subclass Actinobacteridae, order Actinomycetales. While many of these cause disease in humans, Streptomyces is notable as a source of antibiotics. Of those Actinobacteria not in Actinomycetales, Gardnerella is one of the most researched.

Classification of Gardnerella is controversial, and MeSH catalogues it as both a gram-positive and gram-negative organism. Genomes of 45 different strains of Actinobacteria from different genera are either already sequenced or underway right now.

Flagellate

Flagellates are organisms with one or more whip-like organelles called flagella. Some cells in animals may be flagellate, for instance the spermatozoa of most phyla. Flowering plants do not produce flagellate cells, but ferns, mosses, green algae, some gymnosperms and other closely related plants do.

Likewise, most fungi do not produce cells with flagellae, but the primitive fungal chytrids do. Many protists take the form of single-celled flagellates.

The word *flagellate* describe a particular construction of eukaryotic organism and its means of motion. The term does not imply any specific relationship or classification of the organisms that possess flagellae. However, the term "flagellate" is included in other terms (such as "dinoflagellate" and "zooflagellate") which often are more formally characterized.

Form and Behaviour

Eukaryotic flagella are supported by microtubules in a characteristic arrangement, with nine fused pairs surrounding two central singlets. These arise from a basal body or kinetosome, with microtubule roots that are an important part of the cell's brain. In some, for instance, they support a cytostome or mouth, where food is ingested.

The flagella often supports hairs, called mastigonemes, or contain rods. Their ultrastructure plays an important role in classifying eukaryotes. In protoctists and microscopic animals, flagella are generally used for propulsion.

They may also be used to create a current that brings in food. In most things, one or more flagella are located at or near the anterior of the cell e.g. Euglena. Often there is one directed forwards and one trailing behind.

Among animals, fungi, and Choanozoa, which make up a group called the opisthokonts, there is a single posterior flagellum. They are from the phylum Mastigophora. They can cause diseases and they can make their own food.

Protist

Protists are a diverse group of eukaryotic microorganisms. Historically, protists were treated as the kingdom Protista, which includes mostly unicellular organisms that do not fit into the other kingdoms, but this group is contested in modern taxonomy. Instead, it is "better regarded as a loose grouping of 30 or 40 disparate phyla with diverse combinations of trophic modes, mechanisms of motility, cell coverings and life cycles."

The protists do not have much in common besides a relatively simple organization—either they are unicellular, or they are multicellular without specialized tissues. This simple cellular organization distinguishes the protists from other eukaryotes, such as fungi, animals and plants. The term *protista* was first used by Ernst Haeckel in 1866. Protists were traditionally subdivided into several groups based on similarities to the "higher" kingdoms: the one-celled animal-like protozoa, the plant-like protophyta (mostly one-celled algae), and the fungus-like slime molds and water molds. Because these groups often overlap, they have been replaced by phylogenetic-based classifications. However, they are still useful as informal names for describing the morphology and ecology of protists. Protists live in almost any environment that contains liquid water. Many protists, such as the algae, are photosynthetic and are vital primary producers in ecosystems, particularly in the ocean as part of the plankton. Other protists, such as the Kinetoplastids and Apicomplexa, are responsible for a range of serious human diseases, such as malaria and sleeping sickness.

Classification

Historical Classifications: The first division of the protists from other organisms came in the 1830s, when the German biologist Georg A. Goldfuss introduced the word *protozoa* to refer to organisms such as ciliates and corals. This group was expanded in 1845 to include all "unicellular animals", such as Foraminifera and amoebae. The formal taxonomic category *Protoctista* was first proposed in the early 1860s by John Hogg, who argued that the protists should include what he saw as primitive unicellular forms of both plants and animals. He defined the Protoctista as a "fourth kingdom of nature", in addition to the then-traditional kingdoms of plants, animals and minerals.

The kingdom of minerals was later removed from taxonomy by Ernst Haeckel, leaving plants, animals, and the protists as a "kingdom of primitive forms". Herbert Copeland resurrected Hogg's label almost a century later, arguing that "Protoctista" literally meant "first established beings", Copeland complained that Haeckel's term *protista* included anucleated microbes such as bacteria. Copeland's use of the term *protoctista* did not. In contrast, Copeland's term included nucleated eukaryotes such as diatoms, green algae and fungi. This classification was the basis for Whittaker's later definition of Fungi, Animalia, Plantae and Protista as the four kingdoms of life.

The kingdom Protista was later modified to separate prokaryotes into the separate kingdom of Monera, leaving the protists as a group of eukaryotic microorganisms. These five kingdoms remained the accepted classification until the development of molecular phylogenetics in the late 20th century, when it became apparent that neither protists nor monera were single groups of related organisms (they were not monophyletic groups).

Modern Classifications

Currently, the term *protist* is used to refer to unicellular eukaryotes that either exist as independent cells, or if they occur in colonies, do not show differentiation into tissues. The term *protozoa* is used to refer to heterotrophic species of protists that do not form filaments. These terms are not used in current taxonomy, and are retained only as convenient ways to refer to these organisms. The taxonomy of protists is still changing. Newer classifications attempt to present monophyletic groups based on ultrastructure, biochemistry, and genetics. Because the protists as a whole are paraphyletic, such systems often split up or abandon the kingdom, instead treating the protist groups as separate lines of eukaryotes.

The recent scheme by Adl *et al.* (2005) is an example that does not bother with formal ranks (phylum, class, etc.) and instead lists organisms in hierarchical lists. This is intended to make the classification more stable in the long term and easier to update. Some of the main groups of protists, which may be treated as phyla, are listed in the taxobox at right. Many are thought to be monophyletic, though there is still uncertainty. For instance, the excavates are probably not monophyletic and the chromalveolates are probably only monophyletic if the haptophytes and cryptomonads are excluded.

Metabolism

Nutrition in some different types of protists is variable. In flagellates, for example, filter feeding may sometimes occur where the flagella find the prey. Other protists can engulf bacteria and digest them internally, by extending their cell membrane around the food material to form a food vacuole.

Reproduction

Some protists reproduce sexually (conjugation), while others reproduce asexually (binary fission). Some species, for example *Plasmodium falciparum*, have extremely complex life cycles that involve multiple forms of the organism, some of which reproduce sexually and others asexually. However, it is unclear how frequently sexual reproduction causes genetic exchange between different strains of *Plasmodium* in nature and most populations of parasitic protists may be clonal lines that rarely exchange genes with other members of their species.

Role as Pathogens

Some protists are significant pathogens of both animals and plants; for example *Plasmodium falciparum*, which causes malaria in humans, and *Phytophthora infestans*, which causes potato blight. A more thorough understanding of protist biology may allow these diseases to be treated more efficiently. Researchers from the Agricultural Research Service are taking advantage of protists as pathogens in an effort to control red imported fire ant (*Solenopsis invicta*) populations in Argentina. With the help of spore-producing protists such as *Kneallhazia solenopsae* the red fire ant populations can be reduced by 53-100%. Researchers have also found a way to infect phorid flies with the protist without harming the flies. This is important because the flies act as a vector to infect the red fire ant population with the pathogenic protist.

Amoeboid

Amoeboids are single-celled life-forms characterized by an irregular shape. "Amoeboid" and "amœba" are often used interchangeably even by biologists, and especially refer to a creature moving by using pseudopodia. Most references to "amoebas" or "amoebae" are to amoeboids in general rather than to the specific genus *Amoeba*. The genus *Amoeba* and amoeboids in general both derive their names from the ancient Greek word for change.

Structure

Amoeboids move using pseudopodia, which are bulges of cytoplasm. Amoebas breathe using their entire cell membrane that is constantly immersed in water. Excess water can cross into the cytosol. Amoebas have a contractile vacuole to expel excess water. Food sources vary in rhizopoda. They may consume bacteria or other protists. Some are detritivores and eat dead organic material. They extend a pair of pseudopodia around food. They fuse to make a food vacuole which then fuses with a lysosome to add digestive chemicals. Undigested food is expelled at the cell membrane. Amoebas use pseudopodia to move and feed. They are powered by flexible microfilaments near the membrane. Microfilaments are at least 50% of the cytoskeleton. The other parts are more stiff and are composed of intermediate filaments and macrotubules. These are not used in amoeboid movement, but are stiff skeletons on which organelles are supported or can move on.

The shells of amoebas are often composed of calcium. The proteins or materials are synthesised in the cell and exported just outside the cell membrane. Amoebas seem to have connections with two phyla of the lineage fungus-like protists.

The two phyla are myxomycota (plasmodial slime molds), and acrasiomycota (cellular slime molds). These two phyla use amoeboid movement in their feeding stage. One is basically a giant multinucleate amoeba, while the other lives solitary until food runs out; in which a colony of these functions as a unit. Myxomycotes use amoeboid gametes, as well.

Diversity

They have appeared in a number of different groups. Some cells in multicellular animals may be amoeboid, for instance human white blood cells, which consume pathogens. Many protists also exist as individual amoeboid cells, or take such a form at some point in their life-cycle.

The most famous such organism is *Amoeba proteus*; the name amoeba is variously used to describe its close relatives, other organisms similar to it, or the amoeboids in general. As amoebas themselves are polyphyletic and subject to some imprecision in definition, the term "amoeboid" does not provide identification of an organism, and is better understood as description of locomotion. When used in the broader sense, the term can include many different groups.

One source includes 97 different genera. Others include far fewer. In older classification systems, amoeboids have been divided into several morphological categories based on the form and structure of the pseudopods. Those where the pseudopods are supported by regular arrays of microtubules are called actinopods, and forms where they are not are called rhizopods, further divided into lobose, filose, and reticulose amoebae. There is also a strange group of giant marine amoeboids, the xenophyophores, that do not fall into any of these categories. More modern classifications are based upon cladistics. It has been stated that most amoeboid are now grouped in Amoebozoa or Rhizaria. However, in contexts where "amoeboid" is defined more loosely, there are many amoeboid species that are in the Excavata clade.

Pathogenic Interactions with Other Organisms

Some amoeboids can infect other organisms pathogenically (causing disease):

- *Entamoeba histolytica* is the cause of amoebiasis, or *amoebic dysentery.*
- *Naegleria fowleri* (the "brain-eating amoeba") is a fresh-water-native species that can be fatal to humans if introduced through the nose.
- *Acanthamoeba* can cause amoebic keratitis and encephalitis in humans.
- *Balamuthia mandrillaris* is the cause of (often fatal) primary amoebic meningoencephalitis.

Pseudopodia

Pseudopods or pseudopodia (singular: pseudopodium) (from the Greek word "fake, false" +"feet") are temporary projections of eukaryotic cells. Cells having this faculty are generally referred to as amoeboids. Pseudopodia extend and contract by the reversable assembly of actin subunits into microfilaments. Filaments near the cell's end interact with myosin which causes contraction. The pseudopodium extends itself until the actin reassembles itself into a network. This is how amoebas move, as well as some cells found in animals, such as white blood cells.

Reproduction

First, the cell surface extends a membrane process, a lamellipodium. Actin polymerizes, and, at its leading edge, the cell forms filaments that subsequently will blend into one another to form networks. It is supposed that actin polymerization is at the origin of the force propelling the cell forwards. Pseudopodia (which translates to "false feet") are temporary cytoplasm-filled projections of the cell wall that certain eukaryotic cells use for motion or for ingesting nutrients. Most cells with this capability are referred to as amoeboids. Generally several pseudopodia arise from the surface of the body (polypodial-Amoeba proteus) or a single pseudopodium may form on the surface of the body (monopodial-E.histolytica).

Pseudopodia are formed by microtubule and filament structures. The cell surface projects a membrane process called the lamellipodium, which is supported inside by filaments that form at the leading edge, turning into networks as they blend together. Cytoplasm flows into the lamelliopdium, forming the pseudopodia.

The functions of pseudopodia include locomotion and the capturing of prey. Pseudopodia are critical in sensing prey that can then be engulfed; the engulfing pseudopodia are called phagocytosis pseudopodia. A common example of this sort of amoeboid cell is the human white blood cell.

Pseudopodia don't all look like amorphous blobs; instead, they can be classified by their distinct appearances. Lobopodia are bulbous and amoebic. Filopodia are slender, sort of football shaped, and are supported largely by microfilaments. Reticulopodia are very complex and bear individual pseudopodia that form irregular nets. Axopodia are the phagocytosis type with long, thin pseudopods supported by complex, microtubule arrays enveloped with cytoplasm, and they respond rapidly to physical contact.

Morphology

Pseudopods can be classified into several varieties according to their appearance:

- Lobopodia are bulbous, short and blunt in form. They are very typical of Amoebozoa. These finger-like, tubular pseudopodia contain both ectoplasm and endoplasm.
- Filopodia are more slender and filiform with pointed ends, consisting mainly of ectoplasm. These formations are supported by microfilaments. This is observed in *Euglypha* and *Lecithium*.
- Reticulopodia, also known as reticulose pseudopods, are complex formations where individual pseudopods are blended together and form irregular nets. The

primary function of reticulopodia, also known as myxopodia, is the ingestion of food, and the secondary function is locomotion. Reticulopods are typical of Foraminifera.

- Axopodia, also known as actinopodia, are thin pseudopods containing complex arrays of microtubules and are enveloped by cytoplasm. Axopodia are mostly responsible for phagocytosis by rapidly retracting in response to physical contacts. They are observed in radiolaria and heliozoa. This supposedly takes a strain on the helix, for after the sensory action has occurred, it then later on dies. Principally, these pseudopodia are food collecting structures.

Ciliate

The ciliates are a group of protozoans characterized by the presence of hair-like organelles called cilia, which are identical in structure to flagella but typically shorter and present in much larger numbers with a different undulating pattern than flagella. Cilia occur in all members of the group (although the peculiar suctoria only have them for part of the life-cycle) and are variously used in swimming, crawling, attachment, feeding, and sensation. The term "Ciliophora" is used in classification as a phylum. Ciliophora can be classified under Protista or Protozoa. The term "Ciliata" is also used, as a class. (However, this latter term can also refer to a type of fish.) Protista classification is rapidly evolving, and it is not rare to encounter these terms used to describe other hierarchical levels.

Ciliates are one of the most important groups of protists, common almost everywhere there is water — in lakes, ponds, oceans, rivers, and soils. Ciliates have many ectosymbiotic and endosymbiotic members, as well as some obligate and opportunistic parasites. Ciliates tend to be large protozoa, a few reach 2 mm in length, and are some of the most complex protozoans in structure.

Cell Structure

Unlike most other eukaryotes, ciliates have two different sorts of nuclei: a small, diploid micronucleus (reproduction), and a large, polyploid macronucleus (general cell regulation).

The latter is generated from the micronucleus by amplification of the genome and heavy editing. Division of the macronucleus occurs by amitosis, the segregation of the chromosomes is by a process whose mechanism is unknown. This process is by no means perfect, and after about 200 generations the cell shows signs of aging.

Periodically the macronuclei must be regenerated from the micronuclei. In most, this occurs during *conjugation*. Here two cells line up, the micronuclei undergo meiosis, some of the haploid daughters are exchanged and then fuse to form new micronuclei and macronuclei. Food vacuoles are formed through phagocytosis and typically follow a particular path through the cell as their contents are digested and broken down via lysosomes so the substances the vacuole contains are then small enough to diffuse through the membrane of the food vacuole into the cell. Anything left in the food vacuole by the time it reaches the cytoproct (anus) is discharged via exocytosis. Most ciliates also have one or more prominent contractile vacuoles, which collect water and expel it from the cell to maintain osmotic pressure, or in some function to maintain ionic balance. These often have a distinctive star-shape, with each point being a collecting tube.

Feeding

Most ciliates feed on smaller organisms (heterotrophic), such as bacteria and algae, and detritus swept into the oral groove (mouth) by modified oral cilia. This usually includes a series of membranelles to the left of the mouth and a paroral membrane to its right, both of which arise from *polykinetids*, groups of many cilia together with associated structures. The food is moved by the cilia through the mouth pore into the gullet, which forms food vacuoles.

This varies considerably, however. Some ciliates are mouthless and feed by absorption, while others are predatory and feed on other protozoa and in particular on other ciliates. This includes the suctoria, which feed through several specialized tentacles.

Reproduction

Ciliates can undergo both asexual and sexual reproduction. Asexual reproduction occurs by binary fission. The micronucleus undergoes by mitosis and the macronucleus elongates and splits in half. Both new cells each obtain a copy of the micronucleus and macronucleus. Sexual reproduction involves conjugation, which involve two cells. After conjugation, the two cells divide, forming four new cells.

Specialized Structures in Cillates

In some forms there are also body polykinetids, for instance, among the spirotrichs where they generally form bristles called *cirri*. More often body cilia are arranged in *mono-* and *dikinetids*, which respectively include one and two kinetosomes (basal bodies), each of which may support a cilium.

These are arranged into rows called *kineties*, which run from the anterior to posterior of the cell. The body and oral kinetids make up the *infraciliature*, an organization unique to the ciliates and important in their classification, and include various fibrils and microtubules involved in coordinating the cilia. The infraciliature is one of the main component of the cell cortex. Another are the *alveoli*, small vesicles under the cell membrane that are packed against it to form a pellicle maintaining the cell's shape, which varies from flexible and contractile to rigid. Numerous mitochondria and extrusomes are also generally present. The presence of alveoli, the structure of the cilia, the form of mitosis and various other details indicate a close relationship between the ciliates, Apicomplexa, and dinoflagellates. These superficially dissimilar groups make up the alveolates.

Fossil Record

Until recently, the oldest ciliate fossils known were tintinnids from the Ordovician Period. In 2007, Li *et al.* published a description of fossil ciliates from the Doushantuo Formation, about 580 million years ago, in the Ediacaran Period. These included two types of tintinnids and a possible ancestral suctorian.

Classification

Subphylum Postciliodesmatophora:

- Class Karyorelictea
- Class Heterotrichea (e.g. *Stentor*)

Subphylum Intramacronucleata

- Class Spirotrichea
  - Subclass Choreotrichia (e.g. *Tintinnidium*)
  - Subclass Oligotrichia (e.g. *Halteria*)
  - Subclass Stichotrichia (e.g. *Stylonychia*)
  - Subclass Hypotrichia (e.g. *Euplotes*)
- Class Litostomatea
  - Subclass Haptoria (e.g. *Didinium*)
  - Subclass Trichostomatia (e.g. *Balantidium*)
- Class Phyllopharyngea
  - Subclass Phyllopharyngia (e.g *Chilodonella uncinata*
  - Subclass Rhynchodia
  - Subclass Chonotrichia
  - Subclass Suctoria (e.g. *Podophrya*)
- Class Nassophorea
- Class Colpodea (e.g. *Colpoda*)
- Class Prostomatea (e.g. *Coleps*)
- Class Oligohymenophorea
  - Subclass Peniculia (e.g. *Paramecium*)
  - Subclass Hymenostomatia (e.g. *Tetrahymena*)
  - Subclass Scuticociliatia
  - Subclass Peritrichia (e.g. *Vorticella*)
  - Subclass Astromatia
  - Subclass Apostomatia
- Class Plagiopylea

Other

- Genus *Cryptocaryon*

Cilium

A cilium (Latin for eyelash; the plural is *cilia*) is an organelle found in eukaryotic cells. Cilia are slender protuberances that project from the much larger cell body. There are two types of cilia: *motile* cilia and *non-motile*, or primary cilia, which typically serve as sensory organelles. In eukaryotes, cilia and flagella together make up a group of organelles known as undulipodia. Eukaryotic cilia are structurally identical to Eukaryotic flagella, although distinctions are sometimes made according to function and/or length.

Types and Distribution

- Cilia are rare in most plants, occurring most notably in cycads.
- Cilia can be divided into primary and motile forms.

Motile Cilia

Larger eukaryotes, such as mammals, have *motile* cilia as well. Motile cilia are usually present on a cell's surface in large numbers and beat in coordinated waves.

- In humans, for example, motile cilia are found in the lining of the trachea (windpipe), where they sweep mucus and dirt out of the lungs.
- In female mammals, the beating of cilia in the Fallopian tubes moves the ovum from the ovary to the uterus.

Ciliates are microscopic organisms that possess *motile* cilia exclusively and use them for either locomotion or to simply move liquid over their surface.

Primary/Immotile Cilium

In humans, *primary* cilia are found on nearly every cell in the body. In comparison to *motile* cilia, non-motile (or *primary*) cilia usually occur one per cell; nearly all mammalian cells have a single non-motile *primary cilium*. In addition, examples of specialized primary cilia can be found in human sensory organs such as the eye and the nose:

- The outer segment of the rod photoreceptor cell in the human eye is connected to its cell body with a specialized non-motile cilium.
- The dendritic knob of the olfactory neuron, where the odorant receptors are located, also contains non-motile cilia (about 10 cilia per dendritic knob).

Although the primary cilium was discovered in 1898, it was largely ignored for a century. Only recently has great progress been made in understanding the function of the primary cilium.

Until the 1990s, the prevailing view of the primary cilium was that it was merely a vestigial organelle, without important function. Recent findings regarding its physiological roles in chemical sensation, signal transduction, and control of cell growth, have led scientists to acknowledge its importance in cell function, with the discovery of its role in diseases not previously recognized to involve the dysgenesis and dysfunction of cilia, such as polycystic kidney disease congenital heart disease, and an emerging group of genetic ciliopathies.

The primary cilium is now known to play an important role in the function of many human organs. The current scientific understanding of primary cilia views them as "sensory cellular antennae that coordinate a large number of cellular signaling pathways, sometimes coupling the signaling to ciliary motility or alternatively to cell division and differentiation."

Structure, Assembly and Maintenance, and Function

"Inside cilia and flagella is a microtubule-based cytoskeleton called the axoneme. The axoneme of primary cilia typically has a ring of nine outer microtubule doublets (called a 9+0 axoneme), and the axoneme of a motile cilium has two central microtubule singlets in addition to the nine outer doublets (called a 9+2 axoneme). The axonemal cytoskeleton acts as a scaffolding for various protein complexes and provides binding sites for molecular motor proteins such as kinesin II, that help carry proteins up and down the microtubules."

The building blocks of the cilia such as tubulins and other partially assembled axonemal proteins are added to the ciliary tips which point away from the cell body. In most species bi-directional motility called intraflagellar transport (IFT) plays an essential role to move these building materials from the cell body to the assembly site.

IFT also carries the disassembled material to be recycled from the ciliary tip back to the cell body. By regulating the equilibrium between these two IFT processes, the length of cilia can be maintained dynamically. The disassembly of the cilia requires the action of the protein kinase Aurora A.

Exceptions where IFT is not present include *Plasmodium falciparum* which is one of the species of *Plasmodium* that cause malaria in humans. In this parasite, cilia assemble in the cytoplasm. At the base of the cilium where the cilia attaches to the cell body is the microtubule organizing center, the basal body.

Some basal body proteins as CEP164, ODF2 and CEP170, regulate the formation and the stability of the cilium. A transition zone between the basal body and the axoneme "serves as a docking station for intraflagellar transport and motor proteins."

"In effect, the [cilium] is a nanomachine composed of perhaps over 600 proteins in molecular complexes, many of which also function independently as nanomachines."

Sensing the Extracellular Environment

"Some epithelial cells are ciliated, and they commonly exist as a sheet of polarised cells forming a tube or tubule with cilia projecting into the lumen."

Primary cilia on epithelial cells provide chemosenzation, thermosenzation and mechanosenzation of the extracellular environment by playing "a sensory role mediating specific signalling cues, including soluble factors in the external cell environment, a secretory role in which a soluble protein is released to have an effect downstream of the fluid flow, and mediation of fluid flow if the cilia are motile."

Chapter 7

Farma Nanotechnology

Problems are Beta Blocker

One reference lists 29 different uses, including high blood pressure, angina, irregular heart rhythms, migraines, prevention of a second heart attack, tremors, alcohol withdrawal, anxiety and glaucoma. In general, a physician can prescribe medications for uses not approved by the Food and Drug Administration, so long as there is scientific literature to support the use. Usually the pharmaceutical company involved will only research and develop a few uses to submit to the FDA to get the medication approved for marketing.

Once the medication wins FDA approval, the pharmaceutical company is only allowed to advertise it for FDA-approved uses. Over time, however, researchers and physicians gain experience using the medication for uses not approved by the FDA. Beta blockers are not FDA approved for use in anxiety or stage fright, but these medications can be prescribed for this purpose because there is scientific literature to support this use.

Safety of Beta Blockers

Only your physician can tell you whether beta blockers are safe for you. Beta blockers can significantly worsen some medical conditions. As a general rule, beta blockers should not be used in asthmatic people. Beta blockers can worsen congestive heart failure, Raynaud's syndrome, and diabetes.

Musicians who take beta blockers often do so in small doses and only on special occasions, so they may experience no side effects at all from their use. However, many side effects have been reported from these medications, including rash, anaphylactic shock (sudden unconsciousness or death), slowed heart rate, low blood pressure, cold extremities, worsening of heart failure, fainting, rapid heart beats, dizziness, fatigue, headache, depression, sleep disturbances, nightmares, hallucinations, short term memory loss, high or low blood sugar, stomach ache, flatulence, constipation, nausea, diarrhea, dry mouth, vomiting, heartburn, bloating, impotence or decreased libido, difficulty urinating, bronchospasm, cough, wheezes, naal stuffiness, joint pain, and muscle cramps.

Beta Blockers Cause Addiction

Beta blockers do not cause addiction—which means physical and psychological dependence on a substance that is beyond the user's control. But beta blockers, when used on a regular basis, can cause physical dependence.

Physical dependence is when the body becomes accustomed to having a medication in the body, thus over time, more medication is required to achieve the same effect. It also means that if the medication is suddenly stopped, the person can experience a rebound in the symptom the medication was being used to treat. Not all medications cause physical dependence. To develop physical dependence, the medication must be used on a chronic basis.

In the case of beta blockers, physical dependence is not usually an issue when they are used for stage fright, as they are used in low doses and are not taken on a chronic basis. When beta blockers are taken on a chronic basis, as for high blood pressure, suddenly stopping them can cause a severe, dangerous elevation in blood pressure that could lead to a stroke or heart attack.

Addiction causes the person who drinks alcohol, abuses a medication, or takes an illegal substance to exhibit substance-seeking behaviour, perhaps even committing criminal acts to obtain it.

What makes this issue confusing is that alcohol, narcotics, cocaine, and amphetamines can cause both physical dependence and addiction, and usually both must be treated at the same time. Beta blockers do cause physical dependence when used on a chronic basis, but it would be highly unusual for them to cause addiction.

Beta Blockers Help in Performance

The answer varies greatly among individuals. Obviously, if you have a medical condition that makes beta blockers dangerous for you, they will not help.

Beta blockers don't make you play better by themselves; they just relieve physical problems resulting from the fight or flight response. On the other hand, some musicians feel that adrenalin helps their performing, giving them an edge that adds intensity to the performance.

Beta blockers have not been shown to directly improve a musician's emotional state, except to the extent that some musicians feel better when their physical problems are relieved. If your performance anxiety shows itself mainly in psychological ways (*e.g.* negative inner voices), beta blockers will not help you.

The scientific studies and articles we looked at clearly show that beta blockers significantly reduce symptoms that can hinder some people's playing. The musicians in the studies said they felt better about their performance after taking beta blockers, and music critics consistently judged their performances to be better.

A note about dry mouth: Anecdotal reports show that some musicians who've tried beta blockers say they don't help dry mouth, or can even make it worse.

In our research, however, we found a study that conclusively showed beta blockers to help dry mouth among brass players.

Beta Blockers for Performance Anxiety

As you might expect, opinions vary widely. Here are what a few well-known musicians have said on the subject: Trevor Wye, flutist and teacher: After saying that beta blockers "should seriously be considered for important occasions," he writes: "Readers who may have some misgivings about the use of alcohol or drugs may care to reflect on the effect of nerves and strain on the body which can cause a great deal of damage over a period of time.

However, one should use caution by understanding the side effects of a given drug and by realizing that an addiction to drugs may occur and that this may therefore not be the ideal approach to the problem of stage fright.

Dealing with Performance Anxiety

Because the fight or flight response occurs only when we perceive danger, it can be avoided or minimized if we can convince ourselves there is nothing to fear. This is the goal of many non-medication approaches.

The physical sensations of fear can also be managed in specific, non-medication ways. Cognitive therapy, the area of psychotherapy that holds that your understanding of a situation can change your emotional response to it, is often sought by people with stage fright.

Hypnosis may also used. If you decide to seek a therapist, try to find one who is experienced in treating anxiety disorders, and especially one who has experience treating performers.

Different Classes of Beta-Blockers and Specific Drugs

Beta-blockers that are used clinically can be divided into two classes:

- Non-selective blockers (block both b_1 and b_2 receptors), or
- Relatively selective b_1 blockers ("cardioselective" beta-blockers).

Some beta-blockers have additional mechanisms besides beta-blockade that contribute to their unique pharmacologic profile. The two classes of beta-blockers along with specific compounds are listed in the following table.

Clinical Development

The key to the IRCT approach is our ability to combine clinical trials conducted by Encode, our wholly-owned clinical trials subsidiary, with our population data. Encode has enrolled over 3,000 patients in over 30 clinical trials.

In an IRCT the trial cohort is enriched by using population-based epidemiological data as well as data on genetic markers, gene-expression patterns and serum biomarkers correlated with a given disease. This enables deCODE's clinical team to select for a trial patients who are susceptible to disease through the pathway targeted by the drug and are thus likely to respond.

Because of the much larger amount of information going into and coming out of an IRCT, we can test therapeutics for major indications more quickly and with significantly smaller cohorts than would normally be required. The results of an IRCT can then be applied to understand and predict the responsiveness of the patient population at large, information that can be used in further clinical development or to bring the drug to market.

As profit and regulatory compliance pressures increase, life science companies are desperately seeking ways to decrease the cost and manage the risk of their clinical trials. The *Clinical Development Strategy and Technology* service provides life science companies with the necessary insight to increase their operational focus and sharpen their study planning and execution skills.

This Health Industry Insights service provides a forward-looking analysis of IT technologies on the critical path and how they are being adopted all along the clinical development value chain. In addition to qualitative assessment, this service includes quantitative market sizing and forecasts to provide a complete picture of the processes and technologies that impact clinical development.

Approach

This service develops comprehensive data and unique analysis through focused topical surveys, secondary research, and insights from industry experts, practitioners, and vendors. To ensure relevance, Health Industry Insights' analysts work with subscribers to identify and prioritize specific topics to be covered in research reports.

Topics Addressed

Throughout the year, this service will address the following topics:

- The convergence of the eClinical ecosystem

- Clinical performance management and metrics; Clinical data management, Clinical trial management
- Critical tools and systems, including EDC, ePRO, CTMS, CDMS, IVR/IXR
- Evolution towards adaptive trials
- Achieving innovation and optimization of clinical trials IT infrastructure
- Cost containment and risk assessment in clinical development
- eClinical technology landscape, including integration with electronic health record (EHR) and other healthcare IT systems
- Pharmacovigilance and drug safety
- Regulatory compliance issues and data integration standards: CDISC, HL7, ICH

Return on Research Investment

Core Benefits for Technology Buyers:

- A clear view into the benefits, as well as the risks, of adopting eClinical technologies
- The ability to select best-in-class solutions, data interchange standards, and technology vendors
- Strategies to maximize your investment in existing technologies

Core Benefits for Technology Solution Providers:

- An assessment of the market opportunity for eClinical technology solutions
- Unbiased insight from key opinion leaders in the industry
- Increased market exposure and competitive intelligence

Standard Subscription Deliverables

Health Industry Insights values building client relationships and strives to foster a partnership with each client through continuous analyst interaction; concise delivery of current, timely, and relevant research; as well as custom research when needed. Each advisory service delivers strategic research covering business and technology issues.

A summary of the basic service deliverables available to subscribers includes the following:

- Analyst inquiry time specific to the advisory service area
- 24 × 7 research access via healthindustry-insights.com
- Discounted admission to specific Health Industry Insights or IDC conferences
- Complimentary forum invitations

Join Our Conversations

Our Insights Communities are designed to enable technology buyers to engage with our industry analysts, share knowledge and best practices, and connect with colleagues.

To participate in our communities, members create a profile to:

- Participate in our discussions in our blogs, forums and polls.
- Learn about and share best practices, tips, tricks and tools related to business/technology.
- Network with colleagues and join groups within each industry.
- Provide feedback to IDC's industry analysts and hear what we're up to.

Related Offerings

Related offerings to our industry-specific research practices include IT Spending Guides-segmented regional and country-level analysis by solution area; Go-To-Market Services (GMS)-a modular set of deliverables to assist clients in extending their market reach, validating their market strategy, and producing more responsive customer-centric marketing campaigns; Global IT Advisor (GITA)-a flexible, cross-disciplinary research service providing granularity in market information and analysis across a wide range of technology areas; and IT Executive Strategies (ITES)-a best-practices research service covering IT governance, portfolio management, vendor procurement, and more. In addition, Health Industry Insights experts are available for individualized consulting engagements to address specific challenges or to customise the results of our syndicated research offerings.

Four trends stand out: continued rapid growth; increased focus on innovation; differentiation of opportunities; and India's increased importance. Like bamboo, Asia's green shoots grow faster than those in the West. IMS Health predicts low single-digit pharma growth for EU and the US, but double-digit growth rates for APAC over the next five years.

With much of their incremental growth revenues coming from APAC, over the next decade, biopharma companies are shifting R&D spend into the region. This is not just to leverage the well-publicized lower costs and faster recruitment benefits, but is also about market access—as placing global trials in APAC speeds up product launches and product penetration.

For us in the CRO space, this is of course very good news. Business for CROs is predicted to grow even faster than sponsor company growth. Why? Because in addition to the shift of R&D spend to APAC, many sponsor companies simply do not have large organizations in the region. Engaging a CRO means sponsors can go straight to a more agile, and flexible outsourcing model without complications of managing legacy and headcount issues. Growth remains the key word for APAC. Global biopharma and CRO companies are increasingly observing their success in APAC as a surrogate measure of their global success. The companies now realise that they cannot succeed globally without being successful in APAC.

Innovation is becoming a buzz word for APAC. Some of this is traditional biopharma innovation generated in the labs—and the world will see more novel compounds appear from companies such as S-Bio in Singapore. The next decade will also see APAC biopharma companies replicating the process and enabling technology innovation, which is seen in the IT and electronic industries. This innovation will act synergistically with Asian advantages for drug development.

A great example is the drug development enabling website, goBalto.com. This site (incidentally set up by a Korean entrepreneur Jae Chung) will enable Next Generation virtual drug development companies to more easily tap into the burgeoning talent in APAC. Imagine the formation of an Asia-enabled drug development machine, which leverages Asia's lower costs and faster project cycle times, to get products much faster and cheaper than in the West. This has the potential to change the global biopharma operating model in the same way that Japanese car manufacturers transformed the auto industry. For global CROs, there's tremendous opportunity to facilitate innovation in APAC by providing a bridge between East and the West.

As the region matures, greater differentiation is becoming apparent. I have always referred to the APAC region as a portfolio of countries, each with different opportunities. For example, the fast, efficient and predictable regulatory environment in Singapore allowing early site initiation, massive population in India and China providing rapid enrolment and Taiwan's pedigree in oncology studies. Now, there is more clarity about what these differences are, and experienced CROs are better able to allocate development projects across different locations to optimize the mix of benefits and minimize downsides.

Expect this trend to continue. Singapore provides a good example. Recognizing the limitations of its relatively small population, it is becoming a regional center-of-excellence in early-phase studies that require small subject numbers but a fast and predictable regulatory environment. Another 'small' populated country, which is building on an already strong reputation in the early-phase space, is Australia. Expect Australia to increase its focus on Health Economics and Outcomes Research (HEOR) work, as there is a tremendous experience base, thanks to the strict requirements of their regulatory and reimbursement environment.

CROs will also become more differentiated. People in APAC still have difficulty in explaining the differences between various CROs. However, the industry's rapid growth in the region will provide real opportunities to build differentiated CROs in the region. CROs will be differentiated on the basis of service offering and leveraging across centers-of-excellence in different locations in APAC. For example, biometrics capability in India and China, regulatory and HEOR consulting in Australia and market access insight in China and Korea.

Differentiation also will occur with certain product related opportunities. APAC has become the center-of-excellence for oncology studies as highlighted in a recent research article. Biosimilars are an area of real opportunity for APAC with companies such as Celltrion in Korea making global impact in this growing sector. The final trend that is increasingly prominent is India's emergence as the hot growth area in drug development. Volumes have been written and many conference-hours spent discussing India's clinical research capabilities, yet the story is still unfolding. India continues to offer new, unanticipated and interesting opportunities. The original drivers of low cost and access to patients continue to be attractive. India is making its regulatory system more robust, stringent and effective, which will further increase the country's appeal.

Those traditional benefits will ensure India to grow rapidly in its drug development industry for many years. India is also the real source of innovation, particularly in how things can be done better and more efficiently. One such idea is reversing the usual talent deployment approach for successful delivery of international projects. Traditional deployment of talent involved placing project leaders in more mature clinical research markets, while India/APAC's participation remained limited to lower level functions in the value chain. However, working towards maturing local talent to higher level functions in a systematic way offers several advantages including enhanced efficiency, 24/7 leadership coverage for projects and a 'can do' mindset that really resonates well with sponsors. For example, local employees at Kendle India love the career opportunities this provides, with the result, we enjoy as one of the lowest turnover rates of the employees in the industry.

Macro level trends for India includes:

- Relatively stable economy and buoyant local biopharma companies and
- Well rounded progress of clinical development functions marked by addition of newer services, and as reflected in recent strategic deals initiated between top 10 multinational companies with Indian giants.

Role of Food and Drug Administration

As a professional couple with one half in the world of human subjects' protection and the other half in the world of academic pulmonary medicine, we read the article by Miller and Shorr and the accompanying editorial by Perkins with great interest. The article, in the April 2002 issue of *CHEST*, elucidated the many challenges faced by all of us who conduct research involving human subjects. The authors have brought into focus the ethical challenges presented to the investigator, the institutional review board (IRB), and the sponsor. Although we agree with the authors regarding the problems that they have identified, we suggest that the responsibility for the problems is not limited to the three groups they have identified. We would like to point out some of the inherent and fundamental problems of industry-sponsored randomized clinical trials as they are currently conducted.

The current practice is that industry (with some input from expert consultants) designs clinical trials, the US Food and Drug Administration (FDA) reviews them, and then the sponsor then seeks sites for these trials. As Miller and Shorr point out, clinical trials have seen a tremendous change in recent times, moving away from academic medical centers and towards private practice settings.

The scrutiny of IRB applications for these trials is different given the context of an independent IRB relative to the academic IRB. There is a potential and very real problem of "IRB shopping" with the obvious benefit that sponsors can have clinical trials conducted, unopposed and without modification. IRBs, both local and independent, are under extreme pressure to review protocols quickly and without requesting modification.

By contrast, there is a strong tradition and clear need for local control of IRBs so that their actions reflect the communities that they serve. However, Miller and Shorr have pointed out the universal nature of the ethical dilemma confronting the investigators, the IRB, and the sponsor.

Obviously, these groups did not meet this ethical challenge in the mometasone trial. In this context, such review standards and, in fact, this level of review would be more appropriate at the level of the FDA. At the present time, the role that the

FDA plays in the risk/benefit analysis for the subject is ambiguous. This realization leads to a variety of troubling questions.

What was the role of the FDA in allowing a study such as the mometasone trial to proceed? In general, what kind of review does an investigational new drug application undergo? Do experts in the area of study review it? Does the FDA consider whether the protocol meets current guidelines for the treatment of the condition under study, or does the FDA see this as the responsibility of the sponsor or the local IRB?

Our local IRB is frequently told by sponsors that we are the only IRB that has issues with a particular protocol. These issues usually relate to our concerns with including a placebo arm, or with withdrawing or withholding proven treatment.

Industry-sponsored clinical trials that do not pass ethical muster or put patients at unjustifiable risk should be stopped at the FDA review. They should be rewritten according to currently accepted guidelines for the treatment of the condition under study, and according to regulations meant to protect any vulnerable subjects who might be included.

At national IRB meetings last year (Public Responsibility in Medicine and Research), Robert Temple, MD, of the FDA suggested that the agency preferred to allow protocols to have broad inclusion and exclusion criteria so as not to limit multicenter trials. He saw narrowing of inclusion and exclusion criteria to make a study safe as the responsibility of the IRB.

At the same IRB meeting, a representative of the pharmaceutical industry was quick to point out that academic IRBs were a problem for industry, and they would prefer that independent IRBs review all multicentered trials. The implications of this latter statement are obvious.

However, the suggestion of the FDA that narrowing criteria should be done at the local IRB feeds this somewhat dysfunctional relationship. The federal government has now begun to realise that IRB shopping is a real phenomenon. Currently, proposed regulations to prohibit IRB shopping are available for public comment. However, prohibiting IRB shopping alone will not resolve the current problems involving the clinical trials industry that lead to conducting trials such as the one reviewed and analysed by Miller and Shorr.

Unquestionably, unethical clinical trials have been funded by sources other than industry, and have been approved by local IRBs, including academic IRBs. National Institutes of Health-funded studies have also failed the risk/benefit analysis. In this time of heightened public awareness of the shortcomings of the human subjects' protection process, we would argue that rigorous risk/benefit analysis should be applied before protocols reach the local IRBs. In this way, we can help prevent local conflicts of interest and ensure that research is conducted at a standard deserving of the public trust.

The establishment of FDA as it exists today grew out of a time early in the century when consumers were victimized by dishonest purveyors of fraudulent potions and compounds that were ineffective, dangerous, or both.

A system of drug regulation was established in the United States that has served us well. Under this system, FDA reviews new drugs to assess their safety and efficacy. In addition, certain types of drugs must be prescribed and dispensed by only licensed health care professionals.

The prescribing requirement is based on the principle that certain drugs have risks of such significance associated with them that they should be administered only under the supervision and recommendation of a "learned intermediary"- that is, a licensed practitioner with the education and training necessary to oversee the administration of potentially harmful drug products.

Similarly, these products may only be dispensed by a licensed professional that can help to assure proper dosing and administration, and can provide important information on the drug's use to patients. The types of unlawful conduct involving online drug sales that FDA has identified are similar to unlawful activities that occur in other sales contexts. Under the Federal Food, Drug, and Cosmetic Act, FDA has the legal authority to take action against:

- The importation, sale, or distribution of an adulterated or misbranded drug;

- The importation, sale, or distribution of an unapproved new drug;
- Illegal promotion of a drug;
- The sale or dispensing of a prescription drug without a valid prescription; and,
- Counterfeit drugs.

When the Internet is used for an illegal sale, FDA, working with the Department of Justice, must establish the same elements of a case, develop the same charges, and take the same actions as it would if another medium, such as a storefront or a magazine, had been used. FDA has investigated and referred cases for criminal prosecution and initiated civil enforcement actions against online sellers of drugs and other FDA-regulated products, particularly sellers of drugs not approved by the Agency.

As will be described later, FDA has significantly expanded its enforcement activities during this past year with regard to online drug sales.

Code of Federal Regulations (CFR)

Code Of Federal Regulations (CFR) The final regulations published in the Federal Register(daily published record of proposed rules, final rules, meeting notices, etc.) are collected in the CFR. The CFR is divided into 50 titles which represent broad areas subject to Federal regulations. The FDA's portion of the CFR interprets the Federal Food, Drug and Cosmetic Act and related statutes.

Section 21 of the CFR contains all regulations pertaining to food and drugs. The regulations document all actions of all drug sponsors that are required under Federal law. 21CFR Part 314-Applications for FDA Approval to Market a New Drug or an Antibiotic Drug.

CDER's Manual of Policies and Procedures (MaPPs). These documents are approved instructions for internal practices and procedures followed by CDER staff to help standardize the new drug review process and other activities. MaPPs define external activities as well. All MaPPs are available for the public to review to get a better understanding of office policies, definitions, staff responsibilities and procedures.

MaPPS of particular interest to NDA applicants include: Review of the Same Supplemental Change to More than One NDA or ANDA in More Than One Review Division. Filing Review Issues. Action Packages for NDAs and Efficacy Supplements. Refusal to Accept Application for Filing From Applicants in Arrears.Drug Application Approval 50 Policy · Requesting and Accepting Non-Archivable Electronic Records for New Drug Applications. Despite billions of dollars, decades of research, and an unparalleled level of international cooperation between research scientists and clinicians, cancer remains a major cause of death worldwide.

Therapies have improved and many forms of cancer are now treatable, but the disease still kills over six million people throughout the world each year. For this reason, cancer research funding through the National Institute of Health (NIH) and the National Cancer Institute (NCI) represents one of the largest expenditures of the United States federal government. This focus on research has led to improved medical treatments for cancer as well as greater understanding of the molecular intricacies of the disease.

Cancer is now considered not simply one disease but rather a multitude of independent disorders that can all result in malignant cellular growth. Unlike other diseases, there is no simple explanation, or even definition, for cancer, and so there is no simple way to treat it. The dream of a cure for cancer, the "magic bullet" to miraculously eliminate the disease, is now considered unrealistic in light of overwhelming complexities.

However, an interdisciplinary approach has emerged offering new hope for cancer therapies. Standard chemotherapy or radiation regimens are nonspecific in their targeting of cancer cells. In contrast, this new field of therapy reflects a growing awareness on the part of the oncology community that the most effective way to treat a tumor is to target the specific molecular mechanisms responsible for its uncontrolled growth.

Research has shown that not all tumors are the same; molecular targeting, the attempt to tailor therapy to the specific abnormalities causing disease, has become one of the most promising areas in cancer biology research. In addition, as a specific application of molecular targeting, pharmaceutical research has combined the basic science of molecular biology and biochemistry with combinatorial chemistry and organic synthesis to create a new field: rational drug design.

Unfortunately, drug development takes time, and it is years, often decades, before a drug that shows promise in the lab will make it into the clinic. Rational drug design, because it narrows the focus on possible therapeutic agents, is a potential way to shorten this lengthy process. The inherent benefits of the approach are highlighted by one of the most exciting developments in cancer biology in recent years.

On May 10, 2001, as a result of overwhelmingly positive clinical trials, the Food and Drug Administration (FDA) approved Novartis Pharmaceutical's Gleevec (imatinib mesylate) for use in the treatment of chronic myelogenous leukemia (CML). In a stunning validation of the value of rational drug design, Gleevec is the first of a wave of compounds with the potential to revolutionize the way we look at cancer.

The Target-Chronic Myelogenous Leukemia

Leukemia is a disease of the white blood cells. While normal white blood cells, or leukocytes, are responsible for maintaining the immune system, the abnormal proliferation of such cells leads to a deadly form of cancer. Chronic myelogenous leukemia (CML) is characterized by the accumulation of immature granulocytes, a specific type of white blood cell.

The disease progresses through three stages, and the specific stage of disease in a patient is crucial in planning treatment. In stage one, also known as chronic phase, there are few if any outward symptoms of disease; many CML patients are unexpectedly diagnosed as the result of blood work following a routine doctor's appointment. After a period of up to several years, chronic phase CML progresses to accelerated phase CML, which is characterized by an increasing ratio of malignant to normal white blood cells.

The final stage of CML is known as blast crisis, and by this point almost a third of blood and bone marrow cells are cancerous. Due to their high numbers, some of these blast cells may metastasize and form tumors in other tissues including the bones or lymph nodes.

Until recently, the most effective form of treatment for CML was the use of a compound called alpha interferon which was often followed by high dose radiation, chemotherapy, and a bone marrow transplant. However, as CML progresses, the efficacy of bone marrow transplants decreases sharply. Transplants done in the accelerated or blast crisis phase have a high rate of failure. Only 15% of patients transplanted in blast crisis are still living after five years (Leukemia Society). Despite the grim prognosis for CML patients, there is one crucial molecular detail of the disease that renders it a promising target for rational drug design.

The highly specific targeting that led to the laboratory development of Gleevec can be traced back to some of the early studies of leukemia. The first piece of the puzzle was uncovered in 1960 when it was discovered that nearly all patients with CML possess an abnormally small chromosome 22. This abnormality was nicknamed the Philadelphia chromosome and has since become known as CML's cytogenetic calling card.

Shortly thereafter it was discovered that chromosome 9 was elongated by approximately the same length as the missing region of chromosome 22; these findings were eventually linked after molecular characterization confirmed a translocation between chromosomes 9 and 22. In the process of this exchange, breakpoint cluster region (bcr) proteins from chromosome 22 become fused with the c-abl oncogene from chromosome 9 thus creating a hybrid bcr-abl gene.

A normal c-abl gene encodes a protein that is highly regulated; the functional enzyme is shuttled between the nucleus and cytoplasm of a cell so that it is only activated when needed. However, under the influence of bcr resulting from bcr-abl fusion, c-abl loses the checkpoints intended to hold it under control.

The enzyme becomes a constitutively active cytoplasmic protein. Studies in mice have shown that the introduction of the bcr-abl fusion protein, regardless of the presence of the Philadelphia chromosome, is sufficient to experimentally induce the development of CML.

The consequences of the bcr-abl abnormality are devastating because of the resulting c-abl enzyme that is produced. This unregulated enzyme, a tyrosine kinase, is the molecular cause of CML. Kinases are enzymes that transfer phosphate groups from ATP to particular residues, in this case tyrosines, of target proteins within the cell. Phosphorylation often serves as a molecular switch to activate or inactivate a range of proteins involved in such vital activities as cellular growth, development, and survival. For this reason, kinases are normally kept under tight control; a runaway kinase can promote the excessive proliferation associated with cancers including CML.

The Gleevec Application

Since abnormal enzymes are implicated in a variety of human cancers, targeting the mutant function should have deleterious affects for the tumor while leaving normal tissue unaffected. As a result of studies highlighting the molecular mechanisms of CML, a search began for compounds to inhibit the runaway *bcr-abl* kinase.

Compound screening at Novartis Pharmaceuticals in the early 1990's identified 2-phenylaminopyrimidines as a general class of kinase inhibitors, and subsequent chemical synthesis was used to design a series of compounds that should have more specific anti-tumor effects. One of the drugs in this class was called STI571.

In vitro studies using STI571 demonstrated that it was a highly effective and specific inhibitor of platelet-derived-growth-factor (PDGF) kinases, c-kit kinases, and c-abl kinases. STI571 acts by blocking the ATP-binding site of these enzymes.

The functional form of the enzyme uses ATP as its source of phosphate groups, thus by blocking the ATP-binding site of the kinase, STI571 prevents the enzyme from becoming active. In the case of bcr-abl kinases, further experiments demonstrated that STI571 inhibited growth in cell lines carrying the bcr-abl mutation.

Animal models produced data indicating an STI571 dose-dependent inhibition of tumor growth although tumors were not completely eliminated. On the basis of promising laboratory data, STI571 began clinical trials in 1998. The first trials were conducted in patients with chronic phase CML who had previously failed interferon therapy.

As with any Phase I clinical trial, the main purpose of the study was to determine the overall safety and acceptable dose ranges of the drug. However, in the midst of safety evaluations, the physicians running the trial noticed that with elevated doses of STI571 an astounding 98% of the patients showed a complete hematologic response-the abnormal Philadelphia chromosome was still detectable, but the elevated levels of circulating white blood cells, the cause of CML symptoms, were gone.

Due to these astonishing results, more expansive Phase II trials were conducted in over 1000 patients representing a range of CML stages, 27 cancer centers, and six countries. In the 532 chronic phase patients, the results of the original trial were confirmed. Only 3% of patients experienced disease progression while a mere 2% did not complete the trial due to drug side effects. In addition, of the 86% of patients on long term STI571 treatment, 75% achieved either complete or major cytogenetic responses (Mauro and Druker).

Not only did blood counts return to normal, but in many cases the molecular causative agent of disease, the Philadelphia chromosome, had become undetectable. These results were also observed in the 233 accelerated phase patients: 91% achieved a hematologic response, 41% had a cytogenetic response, and mean survival at one year was 74% (Mauro and Druker).

In blast crisis patients, typically the most difficult group to treat, 64% of the 260 patients achieved either a hematologic or cytogenetic response while the median survival of 6.8 months compared favourably with the 3 month median survival for bone marrow transplant recipients (Mauro and Druker).

As a result of its clear clinical benefits for the treatment of CML, in May, after a mere three months of review, the FDA approved STI571 for general use under the trade name of Gleevec. The speed of this decision reflects the amazing significance of Gleevec's application for the treatment of an often fatal disease.

Nanotechnology in Drug Metabolism

The vast majority of drugs are small molecules designed to bind, interact, and modulate the activity of specific biological receptors. Receptors are proteins that bind and interact with other molecules to perform the numerous functions required for the maintenance of life.

They include an immense array of cell-surface receptors (hormone receptors, cell-signaling receptors, neurotransmitter receptors, etc.), enzymes, and other functional proteins.

Due to genetic abnormalities, physiologic stressors, or some combination thereof, the function of specific receptors and enzymes may become altered to the point that our well-being is diminished.

These alterations may manifest as minor physical symptoms, as in the case of a runny nose due to allergies, or as life threatening and debilitating events, such as sepsis or depression. The role of drugs is to correct the functioning of these receptors to remedy the resulting medical condition.

As an example, the highest grossing drug was Prilosec, which earned 4.102 billion in sales. Prilosec is used to treat stomach ulcers and acid reflux disease. Prilosec targets a specific enzyme, the proton pump, which is located in the acid producing cells lining the stomach wall.

This enzyme is responsible for the production of stomach acid. Due to genetic reasons, such as deficient enzymes that regulate acid secretion, or physiologic causes, such as stress, too much acid may be produced. This leads to ulceration of the stomach lining or acid reflux disease and heartburn. Prilosec binds to the proton pump and shuts it down, thereby diminishing the production of stomach acid and its associated symptoms.

Biochemistry of Drugs

Enzymes are a subset of receptor-like proteins that are directly responsible for catalyzing the biochemical reactions that sustain life. For example, digestive enzymes act to break down the nutrients of our diet. DNA polymerase and related enzymes are crucial for cell division and replication.

Enzymes are genetically programmed to be absolutely specific for their appropriate molecular targets. Any errors could have grave consequences. One can imagine the end result should blood clotting enzymes start activating throughout the body. Or consider the problems that arise when our immune system attacks our own tissues.

Enzymes ensure the specificity of their targets by forming a molecular environment that excludes interactions with inappropriate molecules. The analogy most often mentioned is that of a lock and key. The enzyme is a molecular lock, which contains a keyhole that exhibits a very specific and consistent size and shape. This molecular keyhole is termed the*active site* of the enzyme and allows interaction with only the appropriate molecular targets.

Just as a typical lock is much bigger than the keyhole, the receptor is usually much larger than the active site. The receptor, as specified by our DNA, is a folded protein whose major purpose is to form and maintain the size and shape of the active site.

The most important concept in drug design is to understand the methods by which the active site of a receptor selectively restricts the binding of inappropriate structures. Any potential molecule that can bind to a receptor is called a *ligand.*

In order for a ligand to bind, it must contain a specific combination of atoms that presents the correct size, shape, and charge composition in order to bind and interact with the receptor. In essence, the ligand must possess the molecular key that binds the receptor lock. This is termed *steric complementarity.*

As is the case with an actual key, if a different molecule varies by even a single atom in the wrong place, it may not fit properly, and will most likely not interact with the receptor. However, the more closely the fit between the ligand and receptor, the more tightly the interaction becomes. Again, keep in mind that this is only a two-dimensional schematic. Both ligand and active site are volumes with complex three-dimensional shape.

In addition to steric complementarity, electrostatic interactions influence ligand binding. Charged receptor atoms often surround the active site, imparting a localized charge is specific regions of the active site. From physics, we know that opposite charges attract while similar charges repel. *Electrostatic complementarity* further restricts the binding of inappropriate molecules since the ligand must contain correctly placed complementary charged atoms for interaction to occur.

The main driving force for ligand and receptor binding is *hydrophobic interaction.* Nearly two-thirds of the body is water, and this aqueous milieu surrounds all our cells. In order for ligand and receptor to interact, there must be a driving force that compels the ligand to leave the water and bind to the receptor. The hydrophobicity of a ligand is what causes this. Hydrophobicity stands for 'water fearing' and is a measure of how 'greasy' a compound is. It can be roughly approximated by the percentage of hydrogen and carbon in the molecule.

This force is easily demonstrated by placing a few drops of oil in a cup of water. The oil is composed of hydrocarbon chains and is highly hydrophobic.

The oil droplets will instantly coalesce into a single globule in order to avoid the water, which is highly polar. Depending upon the size of the active site, there may be a myriad steric, electrostatic, and hydrophobic contacts. However, some are more important than others.

The specific interactions that are crucial for ligand recognition and binding by the receptor are termed the *pharmacophore.* Usually, these are the interactions that directly factor into the structural integrity of a receptor or are involved in the mechanism of its action. Using our lock and key analogy from above, we can imagine a lock having numerous tumblers.

There may be many keys that can sterically complement the lock and fit within the keyhole. However, all but the correct key will displace the wrong tumblers, leading to a sub-optimal interaction with the lock. Only the correct key, which presents the pharmacophore to the receptor, contacts the appropriate tumblers and properly interacts with the lock to open it. This is crucial to the design of pharmaceuticals since any successful drug must incorporate the appropriate chemical structures and present the pharmacophore to the receptor.

We see our native ligand bound within the active site. Assume that through biochemical investigation, we determine that the phenyl ring and the carboxylic acid group are vital to receptor interaction. Thus, we deduce that these two groups must be the pharmacophore that a ligand must present to the receptor for binding. In future drugs that we develop to mimic the native ligand, we must include these two pharmacophoric elements for successful binding to occur.

This is shown in the upper right derivative compound where a bicyclic group has been substituted. Because it maintains the pharmacophore and retains its complementary size and shape, it has a reasonable chance of successfully binding. However, any drug that we develop which lacks a complete pharmacophore may not interact withz the receptor target.

When a medical condition exists where a drug could be beneficial, extensive scientific study must first be done in order to determine the biological and biochemical problems that underlie the disease process. This often takes years of study in order to characterize the targets for a potential drug. The reason is that nearly all biological processes in the human body are tightly interconnected. Altering the behaviour of select receptors or enzymes may have detrimental effects with other systems.

These are the side effects that occur with nearly all drugs. Furthermore, the human body is a homeostatic machine, and always attempts to achieve equilibrium. As a result, the body will attempt to counteract any pharmacotherapeutic intervention. Once a receptor target has been established and well characterized, the process of ligand design begins. Obviously, the first consideration is that the designed ligand must complement the active site of the receptor target. Steric, electrostatic, and hydrophobic complementarity must be established as we discussed above. The pharmacophore must be presented to the receptor in order for recognition and binding to occur.

In addition to adequately binding the receptor, the biochemical mechanism of the receptor target must be taken into consideration. If the peptide bond is present at a specific position in the active site when the ligand binds, it is cleaved by the protease with the addition of water (H_2O) to form two separate fragments. If our goal is to inactivate this protease, any designed ligand cannot possess this peptide bond at the same position.

Otherwise, it will simply be cleaved by the protease, and the protease will continue to function unperturbed. However, the ligand can be modified so that the peptide bond is no longer present as shown in the bottom portion. If the enzyme then binds this ligand, the enzyme will not be able to cleave it. As such, the enzyme would be inactivated, as the ligand remains lodged in the active site. Having characterized the active site region and the mechanism of action of the target receptor, the challenge then becomes one of designing a suitable ligand. This is, by far, the most daunting task of the entire drug design process.

The optimal combination of atoms and functional groups to complement the receptor is often the natural ligand of the receptor. Unfortunately, this is usually an unacceptable candidate for a drug. This is because the natural ligand either fails to inactivate the receptor, as described above, or it is a natural substance that cannot be patented. Patent considerations are often paramount, as legal protection for the developed drug affords the opportunity to recoup the financial costs of development. Therefore, alternate combinations of chemical structures must be devised.

The design of novel ligands is often restricted by what chemists are physically able to synthesize. It is of no use to design the ultimate drug if it cannot be manufactured. The laws of chemistry dictate that each atom type has a specific size, charge, and geometry with respect to the number and types of neighbouring atoms that it can be joined to. The entire field of chemistry is predicated on the establishment of synthetic rules for the construction and manipulation of various combinations of atoms and functional groups. It is the expertise in these chemical rules that govern the ability of the synthetic chemist to design and synthesize postulated ligand candidates. Within these rules, the drug developer must creatively propose suitable chemical structures that satisfy the requirements discussed above.

Finally, there are biological considerations to the development of new drugs. The liver is the major organ of detoxification in the human body. Any drug that is taken undergoes a number of chemical reactions in the liver as the body attempts to neutralize foreign substances. This set of reactions is well characterized, and a great deal of knowledge exists as to how drugs are modified as the body eliminates them. More importantly, various chemical structures are highly toxic to biological systems, and these are also well characterized. These constraints must also be taken under consideration as novel drugs are developed.

Nanotechnology in Drug Discovery

The result is the determination of specific receptor targets that must be modulated to alter their activity in some way. Once these targets have been identified, the goal is then to find compounds that will interact with the receptors in some fashion. At this initial stage of drug development, it does not matter what effect the compounds have on the targets. We simply wish to find anything that binds to the receptor in any fashion.

The first step is to determine an *assay* for the receptor. An assay is a chemical or biological test that turns positive when a suitable binding agent interacts with the receptor. Usually, this test is some form of colorimetric assay, in which an indicator turns a specific colour when complementary ligands are present.

This assay is then used in *mass screening*, which is a technique whereby hundreds of thousands of compounds can be tested in a matter of days to weeks.

A pharmaceutical company will first screen their entire corporate database of known compounds. The reason is that if a successful match is found, the database compound is usually very well characterized.

Furthermore, synthetic methods will be known for this compound, and patent protection is often present. This enables the company to rapidly prototype a candidate ligand whose chemistry is well known and within the intellectual property of the company.

If a successful match is found, the initial hit is called a *lead compound*. The lead compound is usually a weakly binding ligand with minimal receptor activity. The binding of this structure to the receptor is then studied to determine the interactions that foster the ligand-receptor association. If the receptor is water soluble, there is a chance that *x-ray crystallographic analysis* can be employed to determine the three-dimensional structure of the ligand bound to the receptor at the atomic level.

This is a very powerful tool for it allows scientists to directly visualize a snapshot of the individual atoms of the ligand as they reside within the receptor. This snapshot is referred to as a *crystal structure* of the ligand-receptor complex. Unfortunately, not all complexes can be analysed in this manner.

However, if a crystal structure can be determined, a strategy can then be developed based upon this characterization to improve and optimize the binding of the lead compound. From this point onward, a cycle of iterative chemical refinement and testing continues until a drug is developed that undergoes clinical trials. The techniques most often used to refine drugs are *combinatorial chemistry* and *structure based design.*

Combinatorial chemistry is a very powerful technique that chemists can employ to aid in the refinement of the lead compound. Combinatorial chemistry is a synthetic tool that enables chemists to rapidly generate thousands of lead compound derivatives for testing. A scaffold is employed that contains a portion of the ligand that remains constant.

Subsite groups (shown in red, green, and blue) are potential sites for derivatization. These subsites are then reacted with combinatorial libraries to generate a multitude of derivative structures, each with different substituent groups.

One can see how a vast number of compounds can be generated as a result of the combinatorial process. If a scaffold contains three derivatization sites and the library contains ten groups per site, theoretically 1000 different combinations are possible.

Structure based design, often called rational drug design, is much more focused than combinatorial chemistry. It involves using the biochemical laws of ligand-receptor association discussed above to postulate ligand refinements to improve binding. For example, we discussed that steric complementarity is vital to tight receptor binding. Using the crystal structure of the complex, we can target regions of the ligand that fit poorly within the active site and postulate chemical changes to improve complementarity with the receptor. In a similar fashion, functional groups on the ligand can be changed in order to augment electrostatic complementarity with the receptor.

However, the danger in altering any portion of the ligand is the effect on the remaining ligand structures. Modifying even a single atom in the middle of the ligand can drastically change the shape of the overall structure. Even though complementarity in one portion of the ligand might be improved by the chemical revision, the overall binding might be severely compromised. This is the difficulty in any ligand refinement procedure.

Computers in Drug Design

There was a great deal of optimism that computer aided drug design would revolutionize the way in which drugs could be developed. The enduring exponential increase in computing power had progressed to the point that rudimentary estimations of ligand receptor complementarity could be performed.

Furthermore, computer graphics technology had achieved the ability to generate vector models of chemical structures and manipulate them in real-time. This offered, for the first time, the ability to interactively study computer models of ligand structures and their binding interactions with a receptor.

Concomitant with the development of this technology was the emergence of the AIDS epidemic.Scientists had isolated the causative agent of AIDS, the HIV-1 virus. Considerable characterization of the viral life cycle provided numerous potential targets for pharmaceutical intervention.

Among them was the HIV-1 protease. This aspartyl protease was an enzyme that was unique to HIV, and absolutely required for the processing and maturation of HIV proteins. Thus, if a drug could be developed to inactivate this protease, the virus would be unable to generate mature infectious particles to sustain the infection. Numerous groups around the world rapidly solved the crystal structure of this enzyme. The mechanism of this enzyme was determined, and the layout of the active site was carefully mapped.

It was known that humans possess similar classes of proteases. Renin is an enzyme secreted by the kidneys that is responsible for initiating a cascade of reactions that regulate blood pressure. It too is an aspartic protease, and ligands that inhibited its function were known.

With the wealth of data from the study of the HIV-1 protease, the hope was that this target could be exploited by computer-aided rational drug design to rapidly generate novel AIDS drugs. Computational chemists believed they could circumvent much of the time and effort required for drug synthesis and testing by simply generating novel compounds using the computer. Testing would be replaced by merely calculating the ligand-receptor binding affinity using the physical laws of chemistry. The concept of generating virtual lead compoundsentirely through computer simulation was termed *denovo design.*

Difficulties Implementing Denovo Design

Many of the worlds largest pharmaceutical firms spent millions of dollars on hardware and software in their endeavor to make *denovo* design a reality. Unfortunately, successes were rare. Except for a few exceptions, *denovo* design was an utter failure, and did not prove to be an effective method to discover lead compounds.

The main reasons were limitations in computing power and the lack of useful software functionality. In scientific computing, accuracy and processing time are always a tradeoff.

Thus, in order to make the calculations run in a finite period of time, a plethora of assumptions, significant approximations, and numerous algorithmic shortcuts had to be utilized. This, in turn, greatly diminished the calculated accuracy of any ligand receptor interaction.

As such, chemists could postulate numerous chemical structures that could potentially complement the active site; however, the calculated binding had no correlation with reality.

This remains the most significant challenge in *denovo* design to this day. Although computers have become exponentially faster, the sheer number of calculations needed to accurately predict the binding of a *denovo* generated ligand to its receptor in a useful timeframe still requires significant approximations.

In *denovo* design, we are attempting to generate a whole ligand from scratch and dock it within the receptor. As stated above, the difficulty lies in predicting how the chemical structure will behave in real life.

A ligand is an inherently flexible structure, and can assume a plethora of different conformations and orientations. The big question remains whether the predicted binding structure will mirror the calculated one. Failure in this endeavor has undermined the utility of *denovo* structure generating software. We will discuss these shortcomings and the technological advances of RACHEL, which attempt to circumvent these deficiencies, in detail below.

The second most significant problem in computer aided *denovo* design is the generation of undesired chemical structures. There are a nearly infinite number of potential combinations of atoms. However, the vast majority of these structures are of no use.

As discussed above, undesired structures are rejected due to toxicity, chemical instability, or synthetic difficulty. Nearly all *denovo* design software packages are plagued by this problem, especially with respect to synthetic feasibility. Thus, although such software can postulate potential complementary ligands, the vast majority of them are worthless.

The end result of these shortcomings was that computer aided *denovo* design soon fell out of favour as a means of generating viable lead compounds. By then there had been a tremendous number of *denovo* software packages released; however, they all suffered these same problems.

Gradually, such programs were shelved and investigators looked to other technologies to aid in their drug development efforts. Drug discovery and development is an intense, lengthy and an interdisciplinary endeavor.

Drug discovery is mostly portrayed as a linear, consecutive process that starts with target and lead discovery, followed by lead optimization and pre-clinical *in vitro* and *in vivo* studies to determine if such compounds satisfy a number of pre-set criteria for initiating clinical development.

For the pharmaceutical industry, the number of years to bring a drug from discovery to market is approximately 12-14 years and costing upto $1.2-$1.4 billion dollars.

Traditionally, drugs were discovered by synthesizing compounds in a time-consuming multi-step processes against a battery of *in vivo* biological screens and further investigating the promising candidates for their pharmacokinetic properties, metabolism and potential toxicity. Such a development process has resulted in high attrition rates with failures attributed to poor pharmacokinetics (39%), lack of efficacy (30%), animal toxicity (11%), adverse effects in humans (10%) and various commercial and miscellaneous factors.

Today, the process of drug discovery has been revolutionized with the advent of genomics, proteomics, bioinformatics and efficient technologies like, combinatorial chemistry, high throughput screening (HTS), virtual screening, *de novo* design, *in vitro, in silico*ADMET screening and structure-based drug design.

Silico Drug Design

In silico methods can help in identifying drug targets via bioinformatics tools. They can also be used to analyse the target structures for possible binding/active sites, generate candidate molecules, check for their drug likeness, dock these molecules with the target, rank them according to their binding affinites, further optimize the molecules to improve binding characteristics. The use of computers and computational methods permeates all aspects of drug discovery today and forms the core of structure-based drug design. High-performance computing, data management software and Internet are facilitating the access of huge amount of data generated and transforming the massive complex biological data into workable knowledge in modern day drug discovery process.

The use of complementary experimental and informatics techniques increases the chance of success in many stages of the discovery process, from the identification of novel targets and elucidation of their functions to the discovery and development of lead compounds with desired properties. Computational tools offer the advantage of delivering new drug candidates more quickly and at a lower cost.

Major roles of computation in drug discovery are:

- Virtual screening and de novo design,
- In silicoADME/T prediction and
- Advanced methods for determining protein-ligand binding.

Signification in Silico Drug Design

As structures of more and more protein targets become available through crystallography, NMR and bioinformatics methods, there is an increasing demand for computational tools that can identify and analyse active sites and suggest potential drug molecules that can bind to these sites specifically. Also to combat life-threatening diseases such as AIDS, Tuberculosis, Malaria etc., a global push is essential.

Millions for Viagra and pennies for the diseases of the poor is the current situation of investment in Pharma R&D. Time and cost required for designing a new drug are immense and at an unacceptable level. According to some estimates it costs about $880 million and 14 years of research to develop a new drug before it is introduced in the market Intervention of computers at some plausible steps is imperative to bring down the cost and time required in the drug discovery process

Structure Based Drug Design

The crystal structure of a ligand bound to a protein provides a detailed insight into the interactions made between the protein and the ligand. Structure designed can be used to identify where the ligand can be changed to modulate the physicochemical and ADME properties of the compound,by showing which parts of the compound are important to affinity and which parts can be altered without affecting the binding.

The equlibrium between target and ligand is governed by the free energy of the complex compared to the free energy of the individual target and ligand. This includes not only the interaction between target and ligand but also the solvation and entropy of the three different species and the energy of the conformation of the free species.

Virtual screening and de novo design play an important role within the pharmaceutical industry in lead discovery process. Virtual screening refers to computational screening of large libraries of chemicals for compounds that complement targets of known structure which could be tested experimentally. Since, the virtual screening takes place in the three-dimensional active site of the target, it is also called as structure-based virtual screening.

De novo design attemps to use the unliganded structure of the protein to generate novel chemical structure that can bind. There are varying algorithms, most of which depend on identifying initial putative sites of interaction that are grown into complete ligands.

Fragment based discovery is based on the premise that most ligands that bind stongly to a protein active site can be considered as a number of smaller fragments or functionalities. Fragmnents are identified by screening a relatively small library of molecule by X-ray crystallography, NMR spectroscopy.These structues of the fragment binding to the protein can be used to design new ligands by adding functionality to the fragments or by incorporating features of the fragment onto existing ligands.

The phrase "drug-like" generally means molecules which contain functional groups and/or have properties consistent with the majority of known drugs. Lead structures are ligands that typically exhibit suboptimal target binding affinity. Studies have shown that there exists a difference between leads and drugs which can be expressed as follows: Lead structures exhibit, on average, less molecular complexity (less molecular weight, less number of rings and rotatable bonds), are less hydrophobic (lower ClogP and LogD74) and have lower polarizability (less CMR). Leads should display the following properties to be considered for further development in the drug discovery process or to be called as "drug-like":

- Relatively simple chemical features, amenable for combinatorial and medicinal chemistry optimization efforts;

- Membership to a well established SAR (structure-activity relationship) series, wherein compounds with similar structures exhibit similar target binding affinity;
- Favourable patent situation; and
- Good ADME (absorption, distribution, metabolism and excretion) properties.

Leads discovered using virtual screening and *de novo* design methodologies needs to be optimized to produce candidates with improved bioavailability and low toxicity.

Studies have indicated that poor pharmacokinetics and toxicity are the most important causes of high attrition-rates in drug development and it has been widely accepted that these areas should be considered as early as possible in the drug discovery process, thus improving the efficiency and cost-effectiveness of the industry. Resolving the pharmacokinetic and toxicological properties of drug candidates remains a key challenge for drug developers.

In silico prediction of drug-likeness at an early stage involves evaluation of various ADMET properties using computational approaches like QSAR or molecular modeling. A number of studies have been performed to find out the properties, which make a drug distinct from other chemicals. Availability of large databases of drug or drug-like molecules, *e.g.* CMC (Comprehensive Medicinal Chemistry), MDDR (MACCS-II Drug Data Report), WDI (World Drug Index) provides useful information about the properties of drugs.

The most influential study of "Lipinski's rule-of-five" identifies several critical properties that should be considered for compounds with oral delivery as concern. A deeper understanding of the relationships between important ADME parameters and molecular structure and properties is needed to develop better *in silico* models to predict ADMET properties.

Some of the ADME properties evaluated using *in silico*models are; intestinal permeability, aqueous solubility, human intestinal absorption, human oral bioavailability, active transport, efflux by P-glycoprotein, blood-brain barrier permeation, plasma protein binding, metabolic stability, interactions with cytochrome P450s and toxicity.

Nanoresearch in Drug

There are two major types of drug design. The first is referred to as ligand-based drug design and the second, structure- based drug design.

Ligand Based

Ligand-based drug design (or indirect drug design) relies on knowledge of other molecules that bind to the biological target of interest. These other molecules may be used to derive a pharmacophorewhich defines the minimum necessary structural characteristics a molecule must possess in order to bind to the target. In other words, a model of the biological target may be built based on the knowledge of what binds to it and this model in turn may be used to design new molecular entities that interact with the target.

Structure Based

Structure-based drug design (or direct drug design) relies on knowledge of the three dimensional structure of the biological target obtained through methods such as x-ray crystallography or NMR spectroscopy. If an experimental structure of a target is not available, it may be possible to create a homology model of the target based on the experimental structure of a related protein.

Using the structure of the biological target, candidate drugs that are predicted to bind with high affinity and selectivity to the target may be designed using interactive graphics and the intuition of a medicinal chemist. Alternatively various automated computational procedures may be used to suggest new drug candidates.

As experimental methods such as X-ray crystallography and NMR develop, the amount of information concerning 3D structures of biomolecular targets has increased dramatically. In parallel, information about the structural dynamics and electronic properties about ligands has also increased. This has encouraged the rapid development of the structure-based drug design. Current methods for structure-based drug design can be divided roughly into two categories. The first category is about "finding" ligands for a given receptor, which is usually referred as database searching. In this case, a large number of potential ligand molecules are screened to find those fitting the binding pocket of the receptor.

This method is usually referred as ligand-based drug design. The key advantage of database searching is that it saves synthetic effort to obtain new lead compounds. Another category of structure-based drug design methods is about "building" ligands, which is usually referred as receptor-based drug design. In this case, ligand molecules are built up within the constraints of the binding pocket by assembling small pieces in a stepwise manner. These pieces can be either atoms or fragments.

The key advantage of such a method is that novel structures, not contained in any database, can be suggested. These techniques are raising much excitement to the drug design community.

Active Site Identification

Active site identification is the first step in this program. It analyses the protein to find the binding pocket, derives key interaction sites within the binding pocket, and then prepares the necessary data for Ligand fragment link. The basic inputs for this step are the 3D structure of the protein and a pre-docked ligand in PDB format, as well as their atomic properties. Both ligand and protein atoms need to be classified and their atomic properties should be defined, basically, into four atomic types:

- *Hydrophobic Atom*: All carbons in hydrocarbon chains or in aromatic groups.
- *H-bond Donor*: Oxygen and nitrogen atoms bonded to hydrogen atom(s).
- *H-bond Acceptor*: Oxygen and sp2 or sp hybridized nitrogen atoms with lone electron pair(s).
- *Polar Atom*: Oxygen and nitrogen atoms that are neither H-bond donor nor H-bond acceptor, sulfur, phosphorus, halogen, metal and carbon atoms bonded to hetero-atom(s).

The space inside the ligand binding region would be studied with virtual probe atoms of the four types above so the chemical environment of all spots in the ligand binding region can be known. Hence we are clear what kind of chemical fragments can be put into their corresponding spots in the ligand binding region of the receptor.

Scoring Method

Structure-based drug design attempts to use the structure of proteins as a basis for designing new ligands by applying accepted principles of molecular recognition. The basic assumption underlying structure-based drug design is that a good ligand molecule should bind tightly to its target. Each component reflects a certain kind of free energy alteration during the binding process between a ligand and its target receptor.

The Master Equation is the linear combination of these components. According to Gibbs free energy equation, the relation between dissociation equilibrium constant, Kd and the components of free energy alternation was built.

The sub models of empirical functions differ due to the consideration of researchers. It has long been a scientific challenge to design the sub models. Depending on the modification of them, the empirical scoring function is improved and continuously consummated

Computer-aided Drug Refinement

It was at this time that the techniques of mass screening and combinatorial chemistry began to gain widespread acceptance and use.

The use of mass screening and combinatorial chemistry allowed researchers to discover lead compounds in a rapid and efficient manner. As such, *denovo* design tools and their associated problems were no longer needed to generate lead structures. One would surmise that computer-aided drug design technology would have soon ceased to exist. On the contrary, it soon became apparent that computational tools were needed that could optimize these lead compounds into potent drugs.

The concept of drug optimization versus *denovo* design is an important one. The difficulty with *denovo* ligand generation is that an entire structure is being created from scratch. The confidence one has of accurately predicting how this structure will interact and bind within a target receptor is shaky at best. In drug optimization, we begin with a lead compound whose bound structure within the receptor has been characterized, most likely through x-ray crystallography.

Subtle modifications are then performed to generate derivative compounds using structure based drug design to improve binding affinity. Because we are making much smaller changes, our faith in the validity of the resulting structures is far greater. These derivatives then undergo testing to determine which modifications improve binding. The structures of the best ligands can then be elucidated to verify the accuracy of the modifications. This refinement process continues iteratively until optimal binding ligands are produced.

Since subtle modifications are being made to a common structure, the predictive ability of ligand refinement software is much higher. This is because the effect of a single chemical modification on ligand-receptor binding is far easier to quantitate than an extreme change. No longer are we trying to determine the binding affinities of drastically different structures. Instead, we are simply determining the rank order of a list of derivative compounds. This greatly increases the confidence that proposed structures will bind in a manner consistent with our understanding. In addition, the act of generating chemical derivatives is highly amenable to computerized automation. Consider the application of targeted structure based combinatorial chemistry as discussed above. Libraries of derivative components are assembled based upon the analysis of the active site. Because of the combinatorial nature of this method, an extremely large number of candidate structures may be possible.

A computer can rapidly generate and predict the binding of all potential derivatives, creating a list of the best potential candidates. In essence, the computer filters all weak-binding compounds, allowing the chemist to focus, synthesize, and test only the most promising ligands. Thus, utilizing computer aided drug design software to aid in the refinement of weak binding lead compounds is the most effective manner in which these tools can be employed. The use of computer modeling to refine structures has become standard practice in modern drug design.

Chapter 8

Nano Particle Synthesis

The Size Differentiation

Beside you, the smaller nanocomputer is a block twice your height, but it's easy to climb up onto it as the tourguide suggests. Gravity is less important on a small scale: even a fly can defy gravity to walk on a ceiling, and an ant can lift what would be a truck to us. At a simulated size of fifty nanometers, gravity counts for nothing. Materials keep their strength, and are just as hard to bend or break, but the weight of an object becomes negligible.

Even without the strength-enhancement that lets you overcome molecular stickiness, you could lift an object with 40 million times your mass-like a person of normal size lifting a box containing a half-dozen fully loaded oil tankers. To simulate this weak gravity, the powersuit cradles your body's weight, making you feel as if you were floating. This is almost like a vacation in an orbital theme park, walking with stickyboots on walls, ceilings, and whatnot, but with no need for antinausea medication. On top of the nanocomputer is a stray protein molecule. This looks like a cluster of grapes and is about the same size. It even feels a bit like a bunch of grapes, soft and loose.

The parts don't fly free like a gas or tumble and wander like a liquid, but they do quiver like gelatin and sometimes flop or twist. It is solid enough, but the folded structure is not as strong as your steel fingers. In the 1990s, people began to build molecular machinery out of proteins, copying biology. It worked, but it's easy to see why they moved on to better materials.

From a simulated pocket, you pull out a simulated magnifying glass and look at the simulated protein. The atoms are almost transparent, but even a close look doesn't reveal a nucleus inside, because it's too small to see.

It would take 1,000 times magnification to be able to see it, even with the head start of being able to see atoms with your naked eye. How could people ever confuse big, plump atoms with tiny specks like nuclei?

Remembering how your steel-strong fingers couldn't press more than a fraction of the way toward the nucleus of an argon atom from the air, it's clear why nuclear fusion is so difficult. In fact, the tourguide said that it would take a real-world projectile over a hundred times faster than a high-powered rifle bullet to penetrate into the atomic core and let two nuclei fuse.

Try as you might, there just isn't anything you could find in the molecular world that could reach into the middle of an atom to meddle with its nucleus. You can't touch it and you can't see it, so you stop squinting though the magnifying glass. Nuclei just aren't of much interest in nanotechnology.

Puzzle Chains

Taking the advice of the tourguide, you grab two molecular knobs on the protein and pull. It resists for a moment, but then a loop comes free, letting other loops flop around more, and the whole structure seems to melt into a writhing coil.

After a bit of pulling and wrestling, the protein's structure becomes obvious: It is a long chain-longer than you are tall, if you could get it straight-and each segment of the chain has one of several kinds of knobs sticking off to the side.

With the multicoloured, glassy-bead portrayal of atoms, the protein chain resembles a flamboyant necklace. This may be decorative, but how does it all go back together? The chain flops and twists and thrashes, and you pull and push and twist, but the original tight, solid packing is lost.

There are more ways to go wrong in folding up the chain than there are in solving Rubik's Cube, and now that the folded structure is gone, it isn't even clear what the result should look like. How did those twentieth-century researchers ever solve the notorious "protein folding problem"? It's a matter of record that they started building protein objects in the late 1980s. This protein molecule won't go back together, so you try to break it. A firm grip and a powerful yank straightens a section a bit, but the chain holds together and snaps back. Though unfolding it was easy, even muscles with the strength of steel-the strength of Superman-can't break the chain itself.

Chemical bonds are amazingly strong, so it's time to cheat again. When you say, "Flimsy world-one second!" while pulling, your hands easily move apart, splitting the chain in two before its strength returns to normal. You've forced a chemical change, but there must be easier ways since chemists do their work without tiny superhands.

While you compare the broken ends, they thrash around and bump together. The third time this happens, the chain rejoins, as strong as before. This is like having snap-together parts, but the snaps are far stronger than welded steel. Modern assembler chemistry usually uses other approaches, but seeing this happen makes the idea of molecular assembly more understandable: Put the right pieces together in the right positions, and they snap together to make a bigger structure.

Remembering the "Whoa!" command, you decide to go back to the properly scaled speed for your size and strength. Saying "Standard settings!," you see the thrashing of the protein chain speed up to hard-to-follow blur.

The Nanomachines

At your feet is a ribbed, ringed cylindrical object about the size of a soup can-not a messy, loosely folded strand like the protein (before it fell apart), but a solid piece of modern nanotechnology.

Picking it up, you can immediately feel how different it is from a protein. In the gear, everything is held in place by bonds as strong as those that strung together the beads of the protein chain. It can't unfold, and you'd have to cheat again to break its perfect symmetry. Like those in the wall of the nanocomputer, its solidly attached atoms vibrate only slightly. There's another gear nearby, so you fit them together and make the atomic teeth mesh, with bumps on one fitting into hollows on the other. They stick together, and the soft, slick atomic surfaces let them roll smoothly. Underfoot is the nanocomputer itself, a huge mechanism built in the same rigid style. Climbing down from it, you can see through the transparent layers of the wall to watch the inner works. An electric motor an arm-span wide spins inside, turning a crank that drives a set of oscillating rods, which in turn drive smaller rods.

This doesn't look like a computer; it looks more like an engineer's fantasy from the nineteenth century. But then, it is an antique design-the tourguide said that the original proposal was a piece of exploratory engineering dating from the mid-1980s, a mechanical design that was superseded by improved electronic designs before anyone had the tools to build even a prototype. This simulation is based on a version built by a hobbyist many years later.

The mechanical nanocomputer may be crude, but it does work, and it's a lot smaller and more efficient than the electronic computers of the early 1990s. It's even somewhat faster. The rods slide back and forth in a blur of motion, blocking and unblocking each other in changing patterns, weaving patterns of logic. This nanocomputer is a stripped-down model with almost no memory, useless by itself.

This computer is a millionth of a meter on a side, but from where you stand, it looks like a blocky building looming over ten stories tall. The tourguide says that it contains over 100 billion atoms and stores as much data as a room full of books. You can see some of the storage system inside: row upon row of racks containing spools of molecular tape somewhat like the protein chain, but with simple bumps representing the 1s and 0s of computer data.

These nanocomputers seem big and crude, but the ground you're now standing on is also a computer-a single chip from 1990, roughly as powerful as the smaller, stripped-down nanocomputer at your side. As you gaze out over the chip, you get a better sense for just how crude things were a few decades ago.

At your feet, on the smallest scale, the chip is an irregular mess. Although the wall of the nanocomputer is pebbly with atomic-scale bumps, the bumps are as regular as tile. The chip's surface, though, is a jumble of lumps and mounds.

This pattern spreads for dozens of paces in all directions, ending in an irregular cliff marking the edge of a single transistor. Beyond, you can see other ridges and plateaus stretching off to the horizon.

These form grand, regular patterns, the circuits of the computer. The horizon-the edge of the chip-is so distant that walking there from the center would (as the tourguide warns) take days. And these vast pieces of landscaping were considered twentieth-century miracles of miniaturization?

Cells and Bodies

Even back then, research in molecular biology had revealed the existence of smaller, more perfect machines such as the protein molecules in cells. A simulated human cell-put here because earlier visitors wanted to see the size comparisons-its on the chip next to the smaller nanocomputer.

The tourguide points out that the simulation cheats a bit at this point, making the cell act as though it were in a watery environment instead of air.

The cell dwarfs the nanocomputer, sprawling across the chip surface and rearing into the sky like a small mountain. Walking the nature trail around its edge would lead across many transistor-plateaus and take about an hour.

A glance is enough to show how different it is from a nanocomputer or a gear: it looks organic, it bulges and curves like a blob of liver, but its surface is shaggy with waving molecular chains.

Walking up to its edge, you can see that the membrane wrapping the cell is fluid (cell walls are for stiff things like plants), and the membrane molecules are in constant motion.

On an impulse, you thrust your arm through the membrane and poke around inside. You can feel many proteins bumping and tumbling around in the cell's interior fluid, and a crisscrossing network of protein cables and beams. Somewhere inside are the molecular machines that made all these proteins, but such bits of machinery are embedded in a roiling, organic mass.

When you pull your arm out, the membrane flows closed behind. The fluid, dynamic structure of the cell is largely self healing. That's what let scientists perform experimental surgery on cells with the old, crude tools of the twentieth century: They didn't need to stitch up the holes they made when they poked around inside.

Even a single human cell is huge and complex. No real thinking being could be as small as you are in the simulation: A simple computer without any memory is twice your height, and the larger nanocomputer, the size of an apartment complex, is no smarter than one of the submoronic computers of 1990.

Not even a bendable finger could be as small as your simulated fingers: in the simulation, your fingers are only one atom wide, leaving no room for the slimmest possible tendon, to say nothing of nerves.

For a last look at the organic world, you gaze out past the horizon and see the image of your own, full-sized thumb holding the chip on which you stand. The bulge of your thumb rises ten times higher than Mount Everest.

Above, filling the sky, is a face looming like the Earth seen from orbit, gazing down. It is your own face, with cheeks the size of continents. The eyes are motionless.

Thinking of the tourguide's data, you remember: The simulation uses the standard mechanical scaling rules, so being 40 million times smaller has made you 40 million times faster. To let you pull free of surfaces, it increased your strength by more than a factor of 100, which increased your speed by more than a factor of 10. So one second in the ordinary world corresponds to over 400 million here in the simulation. It would take years to see that huge face in the sky complete a single eyeblink. Enough. At the command "Get me out!", the molecular world vanishes, and your feeling of weight returns as the suit goes slack. You strip off the video goggles-and hugely, slowly, blink.

Bottom-Up Technology

In the previous chapter we read about the sizes, forces, and general nature of objects in the molecular world. Building on this, we can get a better picture of where developments seem to be leading, a better picture of molecular manufacturing itself.

To show the sizes, forces, and general nature of things in molecular manufacturing, we first invite the reader (and the reader's inquisitive alter ego) to take a second and final tour before returning to the world of present-day research. As before, the pre-1990 history is accurate, and the science isn't fiction.

Silicon Valley

The tour of the molecular world showed some products of molecular manufacturing, but didn't show how they were made. The technologies you remember from the old days have mostly been replaced-but how did this happen?

The Silicon Valley Faire is advertised as "An authentic theme park capturing life, work, and play in the early Breakthrough years." Since "work" must include manufacturing, it seems worth a visit.

A broad dome caps the park-"To fully capture the authentic sights, sounds, and smells of the era," the tourguide politely says. Inside, the clothes and hairstyles, the newspaper headlines, the bumper-to-bumper traffic, all look much as they did before your long nap. A light haze obscures the buildings on the far side of the dome, your eyes burn slightly, and the air smells truly authentic.

The Pocket Libraries

The Nanofabricators, Inc., plant offers the main display of early nanotechnology. As you near the building, the tourguide mentions that this is indeed the original manufacturing plant, given landmark status over twenty years ago, then made the centerpiece of the Silicon Valley Faire ten years later, when... With a few taps, you reset the pocket tourguide to speak up less often.

As people file into the Nanofabricator plant, there's a moment of hushed quiet, a sense of walking into history. Nanofabricators: home of the SuperChip, the first mass-market product of nanotechnology. It was the huge memory capacity of SuperChips that made possible the first Pocket Library.

This section of the plant now houses a series of displays, including working replicas of early products. Picking up a Pocket Library, you find that it's not only the size of a wallet, but about the same weight.

Yet it has enough memory to record every volume in the Library of Congress-something like a million times the capacity of a personal computer from 1990. It opens with a flip, the two-panel screen lights up, and a world of written knowledge is at your fingertips.

"Wow, can you believe these things?" says another tourist as he fingers a Pocket Library. "Hardly any video, no 3-D-just words, sound, and flat pictures. And the cost! I wouldn't've bought `em for my kids at that price!" Your tourguide quietly states the price: about what you remember for a top-of-the-line TV set from 1990.

This isn't the cheap manufacturing promised by mature nanotechnology, but it seems like a pretty good price for a library. Hmm... how did they work out the copyrights and royalties? There's a lot more to this product than just the technology...

Nanofabrication

The next room displays more technology. Here in the workroom where SuperChips were first made, early nanotech manufacturing is spread out on display. The whole setup is surprisingly quiet and ordinary.

Back in the 1980s and 1990s, chip plants had carefully controlled clean rooms with gowns and masks on workers and visitors, special workstations, and carefully crafted air flows to keep dust away from products. This room has none of that. It's even a little grubby.

In the middle of a big square table are a half-dozen steel tanks, about the size and shape of old-fashioned milk cans. Each can has a different label identifying its contents: MEMORY BLOCKS, DATA-TRANSMISSION BLOCKS, INTERFACE BLOCKS. These are the parts needed for building up the chip.

Clear plastic tubes, carrying clear and tea-coloured liquids, emerge from the mouths of the milk cans and drape across the table. The tubes end in fist-sized boxes mounted above shallow dishes sitting in a ring around the cans.

As the different liquids drip into each dish, a beater like a kitchen mixer swirls the liquid. In each dish, nanomachines are building SuperChips.

A Nanofab "engineer," dressed in period clothing complete with name badge, is setting up a dish to begin building a new chip. "This," he says, holding up a blank with a pair of tweezers, "is a silicon chip like the ones made with pre-breakthrough technology.

Companies here in this valley made chips like these by melting silicon, freezing it into lumps, sawing the lumps into slices, polishing the slices, and then going through a long series of chemical and photographic steps. When they were done, they had a pattern of lines and blobs of different materials on the surface.

Even the smallest of these blobs contained billions of atoms, and it took several blobs working together to store a single bit of information. A chip this size, the size of your fingernail, could store only a fraction of a billion bits. Here at Nanofab, we used bare silicon chips as a base for building up nanomemory.

The picture on the wall here shows the surface of a blank chip: no transistors, no memory circuits, just fine wires to connect up with the nanomemory we built on top. The nanomemory, even in the early days, stored thousands of billions of bits. And we made them like this, but a thousand at a time-" He places the chip in the dish, presses a button, and the dish begins to fill with liquid.

"A few years latter," he adds, "we got rid of the silicon chips entirely"-he props up a sign saying THIS CHIP BUILD BEGAN AT: 2:15 P.M., ESTIMATED COMPLETION TIME: 1:00 A.M.-" and we sped up the construction process by a factor of a thousand."

The chips in the dishes all look pretty much the same except for colour. The new chip looks like dull metal. The only difference you can see in the older chips, further along in the process, is a smooth rectangular patch covered by a film of darker material.

An animated flowchart on the wall shows how layer upon layer of nanomemory building blocks are grabbed from solution and laid down on the surface to make that film. The tourguide explains that the energy for this process, like the energy for molecular machines within cells, comes from dissolved chemicals-from oxygen and fuel molecules.

The total amount of energy needed here is trivial, because the amount of product is trivial: at the end of the process, the total thickness of nanomemory structure-the memory store for a Pocket Library-amounts to one-tenth the thickness of a sheet of paper, spread over an area smaller than a postage stamp.

The Assembly of Molecules

The animated flowchart showed nanomemory building blocks as big things containing about a hundred thousand atoms apiece (it takes a moment to remember that this is still submicroscopic). The build process in the dishes stacked these blocks to make the memory film on the SuperChip, but how were the blocks themselves built?

The hard part in this molecular-manufacturing business has got to be at the bottom of the whole process, at the stage where molecules are put together to make large, complex parts.

The Silicon Valley Faire offers simulations of this molecular assembly process, and at no extra charge. From the tourguide, you learn that modern assembly processes are complex; that earlier processes-like those used by Nanofabricators, Inc.-used clever-but-obscure engineering tricks; and that the simplest, earliest concepts were never built.

Why not begin at the beginning? A short walk takes you to the Museum of Antique Concepts, the first wing of the Museum of Molecular Manufacturing. A peek inside the first hall shows several people strolling around wearing loosely fitting jumpsuits with attached goggles and gloves, staring at nothing and playing mime with invisible objects.

Oh well, why not join the fools' parade? Stepping through the doorway while wearing the suit is entirely different. The goggles show a normal world outside the door and a molecular world inside.

Now you, too, can see and feel the exhibit that fills the hall. It's much like the earlier simulated molecular world: it shares the standard settings for size, strength, and speed. Again, atoms seem 40 million times larger, about the size of your fingertips. This simulation is a bit less thorough than the last was-you can feel simulated objects, but only with your gloved hands. Again, everything seems to be made of quivering masses of fused marbles, each an atom.

"Welcome," says the tourguide, "to a 1990 concept for a molecular-manufacturing plant. These exploratory engineering designs were never intended for actual use, yet they demonstrate the basics of molecular manufacturing: making parts, testing them, and assembling them."

Machinery fills the hall. Overall, the sight is reminiscent of an automated factory of the 1980s or 1990s. It seems clear enough what must be going on: Big machines stand beside a conveyor belt loaded with half-finished-looking blocks of some material; the machines must do some sort of work on the blocks. Judging by the conveyor belt, the blocks eventually move from one arm to the next until they turn a corner and enter the next hall.

A factory-large enough to make over 10 million nanocomputers per day would fit on the edge one of today's integrated circuits. Inset shows an assembler arm together with workpiece on a conveyor belt.

Since nothing is real, the exhibit can't be damaged, so you walk up to a machine and give it a poke. It seems as solid as the wall of the nanocomputer in the previous tour.

Suddenly, you notice something odd: no bombarding air molecules and no droplets of water-in fact, no loose molecules anywhere. Every atom seems to be part of a mechanical system, quivering thermal vibration, but otherwise perfectly controlled.

Everything here is like the nanocomputer or like the tough little gear; none of it resembles the loosely coiled protein or the roiling mass of the living cell. The conveyor belt seems motionless. At regular intervals along the belt are blocks of material under construction: workpieces. The nearest block is about a hundred marble-bumps wide, so it must contain something like 100 × 100 × 100 atoms, a full million.

This block looks strangely familiar, with its rods, crank, and the rest. It's a nanocomputer-or rather, a blocklike part of a nanocomputer still under construction.

Standing alongside the pieces of nanocomputer on the conveyor belt, dominating the hall, is a row of huge mechanisms.

Their trunks rise from the floor, as thick as old oaks. Even though they bend over, they rear overhead. "Each machine," your tourguide says, "is the arm of a general-purpose molecular assembler.

One assembler arm is bent over with its tip pressed to a block on the conveyor belt. Walking closer, you see molecular assembly in action. The arm ends in a fist-sized knob with a few protruding marbles, like knuckles.

Right now, two quivering marbles-atoms-are pressed into a small hollow in the block. As you watch, the two spheres shift, snapping into place in the block with a quick twitch of motion: a chemical reaction. The assembler arm just stands there, nearly motionless.

The fist has lost two knuckles, and the block of nanocomputer is two atoms larger. The tourguide holds forth: "This general-purpose assembler concept resembles, in essence, the factory robots of the 1980s. It is a computer-controlled mechanical arm that moves molecular tools according to a series of instructions.

Each tool is like a single-shot stapler or rivet gun. It has a handle for the assembler to grab and comes loaded with a little bit of matter-a few atoms-which it attaches to the workpiece by a chemical reaction." This is like the rejoining of the protein chain in the earlier tour.

The Molecular Precision

The atoms seemed to jump into place easily enough; can they jump out of place just as easily? By now the assembler arm has crept back from the surface, leaving a small gap, so you can reach in and poke at the newly added atoms. Poking and prying do no good: When you push as hard as you can (with your simulated fingers as strong as steel), the atoms don't budge by a visible amount. Strong molecular bonds hold them in place.

Your pocket tourguide-which has been applying the power of a thousand 1990s supercomputers to the task of deciding when to speak up-remarks, "Molecular bonds hold things together. In strong, stable materials atoms are either bonded, or they aren't, with no possibilities in between. Assemblers work by making and breaking bonds, so each step either succeeds perfectly or fails completely. In pre-breakthrough manufacturing, parts were always made and put together with small inaccuracies.

These could add up to wreck product quality. At the molecular scale, these problems vanish. Since each step is perfectly precise, little errors can't add up. The process either works, or it doesn't."

But what about those definite, complete failures? Fired by scientific curiosity, you walk to the next assembler, grab the tip, and shake it. Almost nothing happens.

When you shove as hard as you can, the tip moves by about one-tenth of an atomic diameter, then springs back. "Thermal vibrations can cause mistakes by causing parts to come together and form bonds in the wrong place," the tourguide remarks. "Thermal vibrations make floppy objects bend further than stiff ones, and so these assembler arms were designed to be thick and stubby to make them very stiff.

Error rates can be kept to one in a trillion, and so small products can be perfectly regular and perfectly identical. Large products can be almost perfect, having just a few atoms out of place." This should mean high reliability.

Oddly, most of the things you've been seeing outside have looked pretty ordinary-not slick, shiny, and perfect, but rough and homey. They must have been manufactured that way, or made by hand. Slick, shiny things must not impress anyone anymore.

The Molecular Robotics

By now, the assembler arm has moved by several atom-widths. Through the translucent sides of the arm you can see that the arm is full of mechanisms: twirling shafts, gears, and large, slowly turning rings that drive the rotation and extension of joints along the trunk. The whole system is a huge, articulated robot arm.

The arm is big because the smallest parts are the size of marbles, and the machinery inside that makes it move and bend has many, many parts. Inside, another mechanism is at work: The arm now ends in a hole, and you can see the old, spent molecular tool being retracted through a tube down the middle.

Patience, patience. Within a few minutes, a new tool is on its way back up the tube. Eventually, it reaches the end. Shafts twirl, gears turn, and clamps lock the tool in position. Other shafts twirl, and the arm slowly leans up against the workpiece again at a new site. Finally, with a twitch of motion, more atoms jump across, and the block is again just a little bit bigger.

The cycle begins again. This huge arm seems amazingly slow, but the standard simulation settings have shifted speeds by a factor of over 400 million. A few minutes of simulation time correspond to less than a millionth of a second of real time, so this stiff, sluggish arm is completing about a million operations per second.

Peering down at the very base of the assembler arm, you can get a glimpse of yet more assembler-arm machinery underneath the floor: Electric motors spin, and a nanocomputer chugs away, rods pumping furiously.

All these rods and gears move quickly, sliding and turning many times for every cycle of the ponderous arm. This seems inefficient; the mechanical vibrations must generate a lot of heat, so the electric motors must draw a lot of power.

Having a computer control each arm is a lot more awkward now than it was in pre-breakthrough years. Back then, a robot arm was big and expensive and a computer was a cheap chip; now the computer is bigger than the arm. There must be a better way-but then, this is the Museum of Antique Concepts.

Building-Blocks into Buildings

Where do the blocks go, once the assemblers have finished with them? Following the conveyor belt past a dozen arms, you stroll to the end of the hall, turn the corner, and find yourself on a balcony overlooking a vaster hall beyond.

Here, just off the conveyor belt, a block sits in a complex fixture. Its parts are moving, and an enormous arm looms over it like a construction crane.

After a moment, the tourguide speaks up and confirms your suspicion: "After manufacturing, each block is tested.

Large arms pick up properly made blocks. In this hall, the larger arms assemble almost a thousand blocks of various kinds to make a complete nanocomputer.

The grand hall has its own conveyor belt, bearing a series of partially completed nanocomputers. Arrayed along this grand belt is a row of grand arms, able to swing to and fro, to reach down to lesser conveyor belts, pluck million-atom blocks from testing stations, and plug them into the grand workpieces, the nanocomputers under construction. The belt runs the length of the hall, and at the end, finished nanocomputers turn a corner-to a yet-grander hall beyond?

After gazing at the final-assembly hall for several minutes, you notice that nothing seems to have moved. Mere patience won't do: at the rate the smaller arms moved in the hall behind you, each block must take months to complete, and the grand block-handling arms are taking full advantage of the leisure this provides. Building a computer, start to finish, might take a terribly long time. Perhaps as long as the blink of an eye.

Molecular assemblers build blocks that go to block assemblers. The block assemblers build computers, which go to system assemblers, which build systems, which-at least one path from molecules to large products seems clear enough. If a car were assembled by normal-sized robots from a thousand pieces, each piece having been assembled by smaller robots from a thousand smaller pieces, and so on, down and down, then only ten levels of assembly process would separate cars from molecules.

Perhaps, around a few more corners and down a few more ever-larger halls, you would see a post-breakthrough car in the making, with unrecognizable engine parts and comfortable seating being snapped together in a century-long process in a hall so vast that the Pacific Ocean would be a puddle in the corner. Just ten steps in size; eight, starting with blocks as big as the ones made in the hall behind you. The molecular world seems closer, viewed this way.

Molecular Processing

Stepping back into that hall, you wonder how the process begins. In every cycle of their sluggish motion, each molecular assembler gets a fresh tool through a tube from somewhere beneath the floor, and that somewhere is where the story of molecular precision begins. And so you ask, "Where do the tools come from?", and the tourguide replies, "You might want to take the elevator to your left."

Stepping out of the elevator and into the basement, you see a wide hall full of small conveyor belts and pulleys; a large pipe runs down the middle.

A plaque on the wall says, "Mechanochemical processing concept, circa 1990." As usual, all the motions seem rather slow, but in this hall everything that seems designed to move is visibly in motion.

The general flow seems to be away from the pipe, through several steps, and then up through the ceiling toward the hall of assemblers above.

After walking over to the pipe, you can see that it is nearly transparent. Inside is a seething chaos of small molecules: the wall of the pipe is the boundary between loose molecules and controlled ones, but the loose molecules are well confined.

In this simulation, your fingertips are like small molecules. No matter how hard you push, there's no way to drive your finger through the wall of the pipe.

Every few paces along the pipe a fitting juts out, a housing with a mechanically driven rotating thing, exposed to the liquid inside the pipe, but also exposed to a belt running over one of the pulleys, embedded in the housing. It's hard to see exactly what is happening.

The tourguide speaks up, saying, "Pockets on the rotor capture single molecules from the liquid in the pipe. Each rotor pocket has a size and shape that fits just one of the several different kinds of molecule in the liquid, so the process is rather selective. Captured molecules are then pushed into the pockets on the belt that's wrapped over the pulley there, then "Enough," you say. Fine, it singles out molecules and sticks them into this maze of machinery. Presumably, the machines can sort the molecules to make sure the right kinds go to the right places. The belts loop back and forth carrying big, knobby masses of molecules. Many of the pulleys-rollers?-press two belts together inside a housing with auxiliary rollers. While you are looking at one of these, the tourguide says, "Each knob on a belt is a mechanochemical-processing device.

When two knobs on different belts are pressed together in the right way, they are designed to transfer molecular fragments from one to another by means of a mechanically forced chemical reaction.

In this way, small molecules are broken down, recombined, and finally joined to molecular tools of the sort used in the assemblers in the hall above. In this device here, the rollers create a pressure equal to the pressure found halfway to the center of the Earth, speeding a reaction that: "Fine, fine," you say. Chemists in the old days managed to make amazingly complex molecules just by mixing different chemicals together in solution in the right order under the right conditions.

Here, molecules can certainly be brought together in the right order, and the conditions are much better controlled.

It stands to reason that this carefully designed maze of pulleys and belts can do a better job of molecule processing than a test tube full of disorganized liquid ever could.

From a liquid, through a sorter, into a mill, and out as tools: this seems to be the story of molecule processing. All the belts are loops, so the machinery just goes around and around, carrying and transforming molecular parts.

Beyond Antiques

This system of belts seems terribly simple and efficient, compared to the ponderous arms driven by frantic computers in the hall above. Why stop with making simple tools? You must have muttered this, because the tourguide speaks up again and says, "The Special-Assembler Exhibit shows another early molecular-manufacturing concept that uses the principles of this molecule-processing system to build large, complex objects.

If a system is building only a single product, there is no need to have computers and flexible arms move parts around. It is far more efficient to build a machine in which everything just moves on belts at a constant speed, adding small parts to larger ones and then bringing the larger ones together as you saw at the end of the hall above."

This does seem like a more sensible way to churn out a lot of identical products, but it sounds like just more of the same.

Gears like fused marbles, belts like coarse beadwork, drive shafts, pulleys, machines and more machines. In a few places, marbles snap into new patterns to prepare a tool or make a product.

Roll, roll, chug, chug, pop, snap, then roll and chug some more. As you leave the simulation hall, you ask, "Is there anything important I've missed in this molecular manufacturing tour?"

The tourguide launches into a list:

> *"Yes-the inner workings of assembler arms, with drive shafts, worm gears, and harmonic drives; the use of Diels-Alder reactions, interfacial free-radial chain reactions, and dative-bond formation to join blocks together in the larger-scale stages of assembly; different kinds of mechanochemical processing for preparing reactive molecular tools; the use of staged-cascade methods in providing feed-molecules of the right kinds with near-perfect reliability; the differences between efficient and inefficient steps in molecular processing; the use of redundancy to ensure reliability in large systems despite sporadic damage; modern methods of building large objects from smaller blocks; modern electronic nanocomputers; modern methods for-"*

"Enough!" you say, and the tourguide falls silent as you pitch it into a recycling bin. A course in molecular manufacturing isn't what you're looking for right now; the general idea seems clear enough.

It's time to take another look at the world on a more normal scale. Houses, roads, buildings, even the landscape looked different out there beyond the Faire dome-less crowded, paved, and plowed than you remember.

But why? The history books (well, they're more than just books) say that molecular manufacturing made a big difference; perhaps now the changes will make more sense. Yes, it's time to leave.

As you toss your goggled, gloved jumpsuit into another bin, a striking dark-haired woman is taking a fresh one from a rack. She wears a jacket emblazoned with the name "Desert Rose NanoManufacturing."

"How'd you like it?" she asks with a smile.

"Pretty amazing," you say.

"Yes," she agrees. "I saw this sim back when I was taking my first molecular-manufacturing class. I swore I'd never design anything so clunky!

This whole setup really brings back the memories-I can't wait to see if it's as crude as I remember." She steps into the simulation hall and closes the door.

Bibliography

Attebery, B.: *Nanoculture: Implications of the New Technoscience*, Intellect Books, Bristol, 2023.

Banki, L.: *Bioassay of Pesticides in the Laboratory,* Akademiai Kiado, Budapest, 2008.

Barnum, Susan R.: *Biotechnology: An Introduction*, Belmont, Thomson/ Brooks/Cole, 2015.

Barrington, E. J. W.: *Biochemistry of Primitive Deuterostomians*, London, Academic Press, 2014.

Bauer MW: *Biotechnology-the Making of a Global Controversy*, Cambridge, Cambridge University Press, 2012.

Bell, Eric T.: *The Magic of Numbers*, New York, McGraw-Hill, 1996.

Betteridge, K.J. : *Embryo Transfer in Farm Animals,* Ottawa, Agriculture Canada, 2007.

Billingham, R.E., and W.K. Silvers. : *Transplantation of Tissues and Cells*, Wistar Inst. Press, Philadelphia, 1991.

Bower, B.: *Fossils may Clarify Mammal Evolution*, Science News, 2014.

Broach, J.R.: *The Molecular Biology of the Yeast*, Cold Spring Harbor, Cold Spring Harbor Laboratory, 2011.

Castro, G.: *Harvard Soil Mechanics Series 87*, Harvard University, Cambridge, Massachusetts, 1999.

Clymer, R. : *Nature's Healing Agents,* PA, U.S.A: Dorrance Co., 1993.

DeGrazia, David: *Animals Rights: A Very Short Introduction*. Oxford: Oxford University Press, 2022.

Devender Pratap Singh : *A Handbook of Beekeeping,* Agrobios, 2016.

Diaz, E.: *Microbial Biodegradation: Genomics and Molecular Biology,* Caister Academic Press, UK, 2018.

Drexler, K. Eric: *Engines of Creation: The Coming Era of Nanotechnology,* Anchor Books, New York, 2006.

Elliott, B N : *Biochemistry and Molecular Biology*, Oxford University Press, Delhi, 2015.

Ensminger, M.E. : *Dairy Cattle Science,* The Interstate Printers & Publishers, Inc., Danville, 2000.

Escobar, Roberto Calle: *Animal Breeding and Production of Camelids,* Lima, Peru, 2014.

Flowerdew, J. R. : *Animals: Their Reproductive Biology and Population Ecology*, Cambridge Univ. Pr., New York, 2007.

Gaskell, G: *Biotechnology - the Making of a Global Controversy*, Cambridge University Press, Cambridge, 2012.

Gay, W.I. : *Methods of Animal Experimentation*, Academic Press, New York, 1995.

Glimell, H.: *Dynamics of the Emerging Field of Nanoscience*, Kluwer, Dordrecht, 2017.

Godthelp, *Animal: Riversleigh. The Story of Animals Reproduction in Ancient Rainforests*, Reed Books, Balgowhah, 2011.

Goodsell, David S.: *Bionanotechnology: Lessons From Nature*, Hoboken, Wiley-Liss, 2014.

Graf, Alfred Byrd: *Advances in Plant Physiology*, Rajat Pub, Delhi, 2018.

Hacker, J.B. : *Nutritional Limits to Animal Production from Pasture,* Farnham Royal: CAB, 2011.

Hagedorn, A.L.: *Animal Breeding,* Crosby Lockwood, 1990.

Hardy B.: *Biology and Agronomy of Forage Arachis*, Cali, International Centre for Tropical Agriculture, 2014.

Harris J.: *Wonderwoman & Superman: the Ethics of Human Biotechnology,* Oxford University Press, Oxford, 2012.

Hessenbruch, A.: *Nanotechnology and the Negotiation of Novelty*, IOS Press, Amsterdam, 2014.

Holtz, Robert D.; William, Kovacs D.: *An Introduction to Geotechnical Engineering*, Prentice Hall, UK, 2012.

Hunter, Graeme K.: *Vital Forces: The Discovery of the Molecular Basis of Life*, Academic Press, San Diego, 2017.

Hutt, Frederick B.: *Genetics for Dog Breeders*, Freeman & Company, 2009.

James Lewis: *Prospects in Nanotechnology: Toward Molecular Manufacturing,* John Wiley & Sons, New York, 2015.

Jonoska, N.: *Self-Assembling DNA Graphs, DNA-Based Computers VIII*, Springer-Verlag, Berlin, 2013.

Joysey, K. A. : *Development, Function and Evolution of Animal Teeth*, Academic Pr., New York, 2008.

Kallenbach, N.R.: *Molecular Structure: Chemical Reactivity and Biological Activity,* Oxford University Press, Oxford, 2008.

Kovach, S.P.: *Injection of Fertilizers into Drip Irrigation Systems for Vegetables,* Univ. Florida, Gainesville, USA, 2004.

Kramer, Steven L.: *Geotechnical Earthquake Engineering,* Prentice Hall, UK, 2016.

Krieger, Maggie and Richard: *Secrets of the Andean Alpaca - The Field Guide,* Saltspring Island Llamas and Alpacas, 2014.

Kunelius T.: *Annual Ryegrasses in Atlantic Canada,* Ottawa, Agriculture Canada, 2011.

Liu W.K., Karpov E.G., Park H.S.: *Nano Mechanics and Materials,* Wiley, Weinheim, 2016.

Lyster, S. : *Animals and Their Moral Standing.* London : Routledge, 2017.

Macself, A.J.: *Soils and Fertilizers,* Satish Serial Pub, Delhi, 2015.

Mahindru, S. N.: *Food Safety and Pesticides,* APH, Delhi, 2019.

Mathialagan, P : *Textbook of Animal Husbandry and Livestock Extension,* International Book Distributing Co, Delhi, 2015.

Matthews, L. H.: *The Life of Animals Reproduction,* London, Weidenfield and Nicholson, 1999.

Meena Francis: *Biotech's Dictionary of Biochemistry,* Biotech Books, Delhi, 2017.

Mondal, R.C.: *Appropri-ate Technology,* Intermediate Technology Publications, London, 2004.

Moravec, Hans: *Mind Children: The Future of Robot and Human Intelligence,* Harvard University Press, Cambridge, 2008.

Narasimha Rao, P.: *Irrigation Development: Issues and Challenges,* Discovery, 2017.

Paul M. Althouse: *Introduction to Agricultural Biochemistry,* Biotech Books, Delhi, 2015.

Pemberton, R. W.: *Predictable Risk to Native Plants in Weed Biological Control,* Oecologia, 2013.

Pressman, J.: *Nano Narrative: A Parable from Electronic Literature,* Intellect Books, Bristol, 2014.

Ratner, Daniel: *Nanotechnology: a Gentle Introduction to the Next Big Idea,* Prentice Hall, Upper Saddle River, 2013.

Raymond, F., Redman, P., & Waltham, R. : *Forage conservation and Feeding.* Ipswich: Farming Press, 2006.

Renaville, R and A Burny : *Biotechnology in Animal Husbandry,* Springer Pub, 2018.

Rietman, Ed.: *Molecular Engineering of Nanosystems,* Springer, New York, 2016.

Roobeek, A: *The Biotechnology Revolution?,* Blackwell, Oxford, 2015.

Safley, Michael: *Synthesis of a Miracle*, Northwest Alpacas, 2022.

Schiller, A. L. : *Anatomy of the Guinea Pig*, Harvard Univ. Pr., Cambridge, 1995.

Schummer, J.: *Interdisciplinary Issues in Nanoscale Research*, IOS Press, Amsterdam, 2014.

Seeman, N.: *DNA Nanotechnology for Massive Information Storage,* New York University, New York, 2014.

Short, R. V.: *Reproduction in Mammals*, Cambridge, Cambridge University Press, 1992.

Stephen M. Goodnick: *Transport in Nanostructures,* Cambridge University Press, Cambridge, 2017.

Suchman, M.: *Envisioning Life on the Nano-Frontier*, Kluwer, Dordrecht, 2011.

Swarnim, K. : *A Textbook of Biochemistry and Microbiology*, Surendra Pub, Delhi, 2020.

Tenner, E.: *Nanotechnology and Unintended Consequences*, Kluwer, Dordrecht, 2011.

Thornhill, Nancy W.: *The Natural History of Inbreeding and Outbreeding*, Chicago Press, 2013.

Whealy K.: *The Garden Seed Inventory*, Decorah, Seed Saver Publications, 2008.

White, M.J.D.: *Animal Cytology and Evolution*, Cambridge, Cambridge Univ. Press, 1994.

Whiting, G.C.: *The Biochemistry of Fruits and their Products*, Academic Press, London & New York, 2000.

Index